2001 | Intermediate 2

[BLANK PAGE]

FOR OFFICIAL USE

X007/201

| | Total for Sections B and C | |

NATIONAL QUALIFICATIONS 2001

MONDAY, 21 MAY 9.00 AM – 11.00 AM

BIOLOGY INTERMEDIATE 2

Fill in these boxes and read what is printed below.

Full name of centre

Town

Forename(s)

Surname

Date of birth
Day Month Year   Scottish candidate number   Number of seat

SECTION A (25 marks)
Instructions for completion of Section A are given on page two.

SECTIONS B AND C (75 marks)

1 (a) All questions should be attempted.
   (b) It should be noted that in **Section C** questions 1 and 2 each contain a choice.

2 The questions may be answered in any order but all answers are to be written in the spaces provided in this answer book, and must be written clearly and legibly in ink.

3 Additional space for answers and rough work will be found at the end of the book. If further space is required, supplementary sheets may be obtained from the invigilator and should be inserted inside the **front** cover of this book.

4 The numbers of questions must be clearly inserted with any answers written in the additional space.

5 Rough work, if any should be necessary, should be written in this book and then scored through when the fair copy has been written.

6 Before leaving the examination room you must give this book to the invigilator. If you do not, you may lose all the marks for this paper.

Read carefully

1 Check that the answer sheet provided is for Biology Intermediate 2 (Section A).

2 Fill in the details required on the answer sheet.

3 In this section a question is answered by indicating the choice A, B, C or D by a stroke made in **ink** in the appropriate place in the answer sheet—see the sample question below.

4 For each question there is only **one** correct answer.

5 Rough working, if required, should be done only on this question paper, or on the rough working sheet provided—**not** on the answer sheet.

6 At the end of the examination the answer sheet for Section A **must** be placed inside the front cover of this answer book.

Sample Question

Which of the following lists all the elements that are present in every protein molecule?

A Carbon, oxygen, nitrogen

B Carbon, hydrogen, oxygen, nitrogen

C Carbon, hydrogen, oxygen, sulphur

D Carbon, hydrogen, oxygen

The correct answer is B—Carbon, hydrogen, oxygen, nitrogen. A **heavy** vertical line should be drawn joining the two dots in the appropriate box in the column headed **B** as shown **in the example on the answer sheet**.

If, after you have recorded your answer, you decide that you have made an error and wish to make a change, you should cancel the original answer and put a vertical stroke in the box you now consider to be correct. Thus, if you want to change an answer **D** to an answer **B**, your answer sheet would look like this:

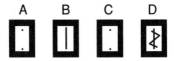

If you want to change back to an answer which has already been scored out, you should **enter a tick (✓)** to the RIGHT of the box of your choice, thus:

SECTION A

All questions in this Section should be attempted.

1. Which term describes all the organisms living in the same area?

   A Food web
   B Ecosystem
   C Population
   D Community

2. Which term describes an animal that eats other animals in a food web?

   A Prey
   B Predator
   C Producer
   D Primary consumer

3. The diagram below shows a pyramid of biomass.

   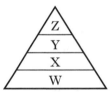

   W represents the total mass of

   A producers
   B prey
   C predators
   D herbivores.

Questions 4, 5 and 6 refer to the diagram and the information below.

A choice chamber was set up to investigate the response of woodlice to an environmental condition.

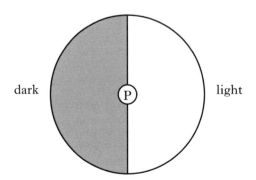

Thirty woodlice were placed in the choice chamber at point P.

The woodlice distribution was noted every 2 minutes during a 10 minute period.

The results are shown in the table below.

Time (minutes)	Number of woodlice	
	In the dark	In the light
2	12	18
4	13	17
6	15	15
8	19	11
10	27	3

4. The environmental condition being investigated is

   A temperature
   B number of woodlice
   C light
   D humidity.

5. The simple whole number ratio of woodlice found in the dark to those found in the light after two minutes was

   A 2 : 3
   B 3 : 2
   C 6 : 9
   D 12 : 18.

6. A possible conclusion from this investigation is

   A woodlice are more likely to be found in the light
   B woodlice are more likely to be found in the dark
   C woodlice cannot detect light
   D woodlice show no difference in their response to light and dark.

7. What must be present in leaf cells for photosynthesis to take place?

   A Carbon dioxide and water

   B Oxygen and water

   C Carbon dioxide and oxygen

   D Oxygen and hydrogen

8. Products of the photolysis stage of photosynthesis are

   A glucose and hydrogen

   B carbon dioxide and hydrogen

   C water and oxygen

   D hydrogen and oxygen.

9. The diagram below shows the carbon fixation stage of photosynthesis.

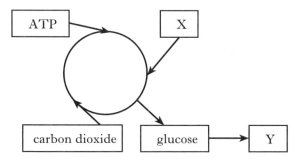

Which line in the table correctly identifies X and Y?

	X	Y
A	hydrogen	starch
B	starch	ADP
C	starch	oxygen
D	water	starch

10. The graph below shows the changing pH of a sample of milk over a seven day period.

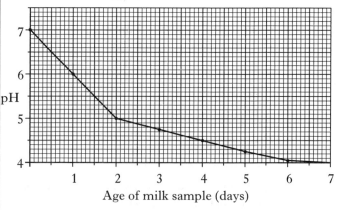

The changes in pH are due to

A lactic acid in the milk destroying the bacteria

B enzymes in the milk being denatured

C bacteria in the milk causing the production of lactic acid

D bacteria in the milk producing carbon dioxide.

Questions 11 and 12 refer to the diagram below which shows a section through a mammalian heart.

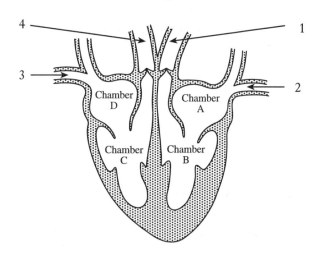

11. Which arrow identifies the vena cava?

    A 1

    B 2

    C 3

    D 4

12. Which heart chamber pumps blood to the lungs?

13. Which letter correctly identifies the location of the sensory strip of the cerebrum?

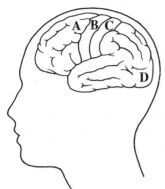

14. The diagram below shows a site of gas exchange in the lungs.

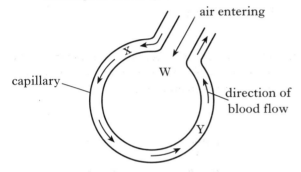

Which line in the table shows the correct relative concentrations of oxygen?

	Concentration of oxygen		
	at W	at X	at Y
A	high	low	high
B	low	high	low
C	low	high	high
D	high	high	low

Questions 15 and 16 refer to the graph below.

The graph shows the percentage saturation of haemoglobin at different oxygen concentrations.

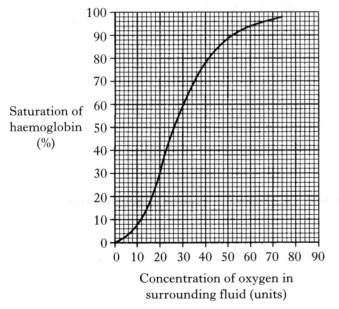

15. What is the percentage saturation of haemoglobin with oxygen when the oxygen concentration of the surroundings is 60 units?

A 30

B 90

C 92

D 94

16. What is the increase in percentage saturation of haemoglobin with oxygen as the oxygen concentration of the surroundings increases from 20 to 40 units?

A 20

B 30

C 48

D 78

17. Digestion takes place in animals

A and allows insoluble molecules to pass directly through the wall of the small intestine

B as enzymes cannot act on insoluble molecules

C and makes insoluble molecules soluble to allow absorption

D and allows food to be passed along the gut by peristalsis.

18. The hepatic portal vein transports

A glucose from the large intestine to the liver

B amino acids from the small intestine to the liver

C fats from the lacteal to the liver

D urea from the kidneys to the liver.

19. Which substances are normally excreted in urine?

A Protein and urea

B Urea and salts

C Glucose and salts

D Protein and salts

[Turn over

20. The table below shows water gained and lost by the body over a 24 hour period.

Method of water gain	Volume of water gained (cm³)	Method of water loss	Volume of water lost (cm³)
food	800	exhaled breath	
drink	1000	sweating	500
metabolic water	350	urine	1250
		faeces	100

What volume of water is lost in exhaled breath?

A 100 cm³
B 200 cm³
C 300 cm³
D 500 cm³

21. After running a race an athlete experienced muscle fatigue.

Which of the following increased in the muscles?

A Glucose
B Oxygen
C ATP
D Lactic acid

22. During aerobic respiration in muscle cells oxygen will combine with

A hydrogen to form water
B pyruvic acid to form lactic acid
C pyruvic acid to form ethanol
D glucose to form pyruvic acid.

23. Which of the following is a correct description of a chromosome?

A A chain of DNA bases
B A chain of RNA bases
C A chain of amino acids
D A chain of sugar molecules

24. If an inherited characteristic is controlled by alleles of more than one gene, then the type of inheritance is called

A true breeding
B polygenic
C co-dominant
D monohybrid.

25. The diagram below represents the transmission of sex determining chromosomes from parents to offspring.

```
        mother            father
        / \                / \
Gametes P   Q            R   S
```

Which line in the table below correctly identifies the sex chromosomes for the gametes P, Q, R and S?

	Gamete P	Gamete Q	Gamete R	Gamete S
A	XX	XX	XY	XY
B	X	X	X	Y
C	XX	XY	XX	XY
D	X	Y	X	Y

Candidates are reminded that the answer sheet for Section A MUST be placed inside the front cover of this answer book.

SECTION B

All questions in this section should be attempted.

1. The diagrams below show sections of three different cell types.
   They are not drawn to the same scale.

   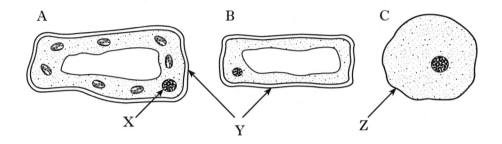

   (a) (i) Complete the table below to show the name and function of the parts labelled X, Y and Z.

Part of cell	Name	Function
X		
Y		
Z		

   (ii) Cells A and B are plant cells and cell C is an animal cell.
        Describe **two** features, shown in the diagrams, that support this statement.

   _____

   _____

   (b) Name a substance produced by cell type A that can be used by cell types B and C.

   _____

   [Turn over

2. (a) An investigation into the effect of temperature on anaerobic respiration in yeast was carried out.

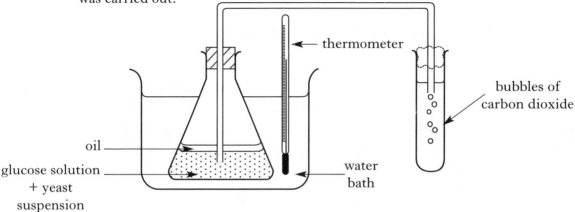

1. A glucose solution was boiled and cooled and poured into a conical flask.
2. A yeast suspension was added to the glucose solution.
3. Oil was poured over the surface of the liquid.
4. The number of bubbles of carbon dioxide produced in one minute was counted.
5. The procedure was repeated at a range of temperatures.

(i) In this investigation temperature was the only variable altered.

State **two** variables that should be kept constant when setting up this investigation.

1 _____

2 _____

(ii) Explain the purpose of the layer of oil.

_____

(iii) The results are shown in the table below.

Temperature (°C)	Bubbles of carbon dioxide (number/minute)
4	0
20	3
35	6
45	22
50	20
70	0

2. (a) (continued)

Present the results in a suitable format on the grid below.
(Additional graph paper, if required, will be found on page 28.)

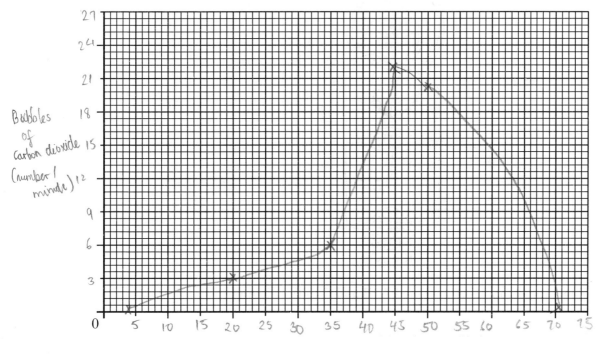

Temperature (°C)

(iv) From the results, describe the effect of temperature on anaerobic respiration in yeast.

_____

_____   2

(v) Suggest **one** way in which the reliability of the results could be improved.

_____   1

(vi) In addition to carbon dioxide, what will be produced in the flask during the investigation?

_____   1

(vii) Explain why no carbon dioxide is produced when the temperature is 70 °C.

_____   1

(b) Yeast is a micro-organism used in the production of bread.
Name **one** other type of micro-organism and an associated product.

_____ used to produce _____.   1

3. The diagram below shows part of the human circulatory system.

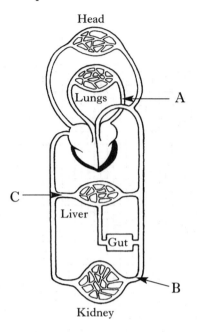

(a) Name the blood vessels labelled A, B and C in the diagram.

A _____

B _____

C _____ 3

(b) The diagram below shows a cross section through an artery.

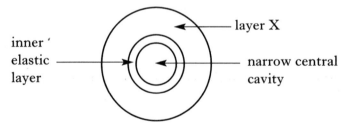

(i) Describe layer X.

_____ 1

(ii) Describe how the structure of layer X is related to its function.

_____ 1

(iii) The pulse beat can be felt in an artery.

What causes the pulse beat?

_____ 1

4. (a) (i) Describe **two** features of the lungs which make them efficient gas exchange surfaces.

1 _____

2 _____

(ii) Name the structures in the lungs through which oxygen diffuses from the air into the blood capillaries.

_____

(b) The table below shows the rates of oxygen intake and energy release during various activities.

Activity	Rate of oxygen intake (cm^3/second)	Rate of energy release (joules/second)
Resting	3·5	70
Playing golf	20	400
Playing tennis	25	500
Playing football	30	600

Describe the relationship between oxygen intake and energy release.

_____

[Turn over

5. (a) Some information on digestion and absorption of the main food groups is given in Table 1 below.

Table 1

Food group	Product(s) of digestion	Use after absorption
Carbohydrate		Energy production
Protein		Making new proteins for growth and repair
Fat	Fatty acids and glycerol	

(i) Complete the table to show the products of digestion and their uses after absorption. **2**

(ii) Name the chemical elements present in fat molecules.

_____ **1**

(b) The nutritional information for 100 g of a breakfast cereal is shown in Table 2 below.

Table 2

	Each 100g of cereal provides
Energy	1440 kJ
Protein	11·2 g
Carbohydrate	67·6 g
Fat	2·7 g
Vitamins	18·0 mg
Fibre	10·5 g

DO NOT WRITE IN THIS MARGIN

Marks

5. (*b*) (continued)

(i) The total daily energy requirement for a sixteen year old girl is 9600 kJ.

Calculate the percentage of her daily total energy requirement that would be provided by a 50 g portion of breakfast cereal each day.

Space for calculation

________________ % **1**

(ii) Which food group is missing from Table 2?

____________________________ **1**

[Turn over

6. (a) Pepsin is an enzyme that breaks down protein.

Photographic film that has been exposed and developed has a black coating that is held on by a protein layer.

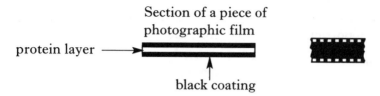

The diagram below shows the results of an investigation into the activity of the enzyme pepsin.

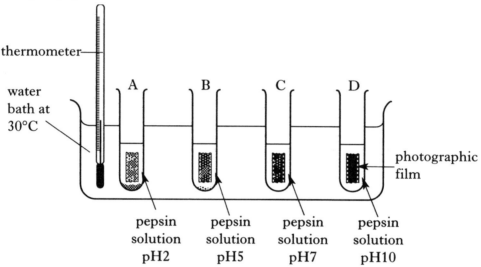

(i) Name the variable studied in this investigation.

_____ 1

(ii) Explain why most protein was broken down in Tube A.

_____ 1

(iii) Describe suitable controls to show that the enzyme is causing the observed effects.

_____
_____ 1

(b) Pepsin is the enzyme produced by secretory cells in the stomach.

Describe the function of the mucus also produced in the stomach.

_____ 1

7. (a) Groups of students carried out an investigation into the effects of competition. Trays containing seeds were set up as shown below.

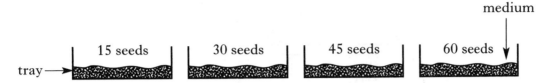

The trays were watered regularly to allow germination to take place.

After several days the seedlings were observed and the number with healthy green leaves was noted.

The results are shown in the table below.

Number of seeds in each tray	Number of seedlings with healthy green leaves	Percentage of seedlings with healthy green leaves
15	12	80
30	18	60
45	23	51
60	24	40

(i) Predict the percentage of seedlings with healthy green leaves if 75 seeds were sown in a seed tray.

_____ 1

(ii) State **two** factors for which the seedlings could be competing in this investigation.

1 _____

2 _____ 1

[Turn over

7. (continued)

(b) A horticulturist grew a crop of peas in a polythene tunnel.

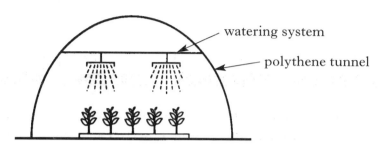

State **two** conditions that could limit the rate of photosynthesis in the pea plants.

1 _____

2 _____   2

(c) The graph below shows the mass of sugar in the leaves and the concentration of carbon dioxide just above the leaves of pea plants at different times of the day.

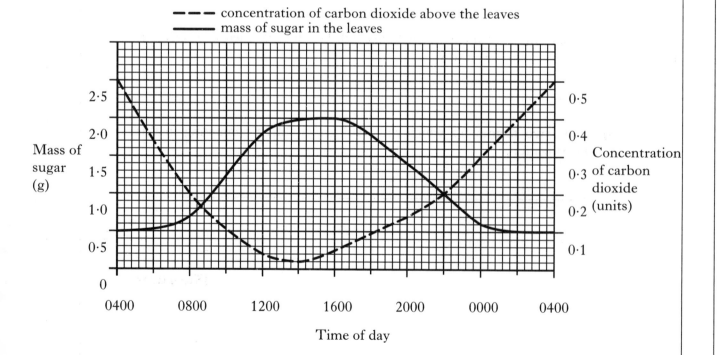

(i) For how many hours did the carbon dioxide concentration decrease?

_____ hours   1

DO NOT WRITE IN THIS MARGIN

Marks

7. (*c*) (continued)

(ii) Explain this decrease in carbon dioxide concentration.

____

____ 1

(iii) Explain why the mass of sugar in the leaves is at its maximum at 1600.

____

____ 1

(iv) Calculate the **percentage increase** in sugar in the leaves between 0400 and 1600.

Space for calculation

__________ % 1

(v) Give **two** reasons why the mass of sugar decreases between 2000 and 0000.

1 ____

2 ____ 2

[Turn over

DO NOT WRITE IN THIS MARGIN

Marks

8. (*a*) Name the male gametes and their site of production in a flower.

Male gametes ______________________________

Site of production ______________________________ **2**

(*b*) Tomato plants were used in an experimental monohybrid cross.

The parental phenotypes were cut leaf and potato leaf as shown in the diagrams below.

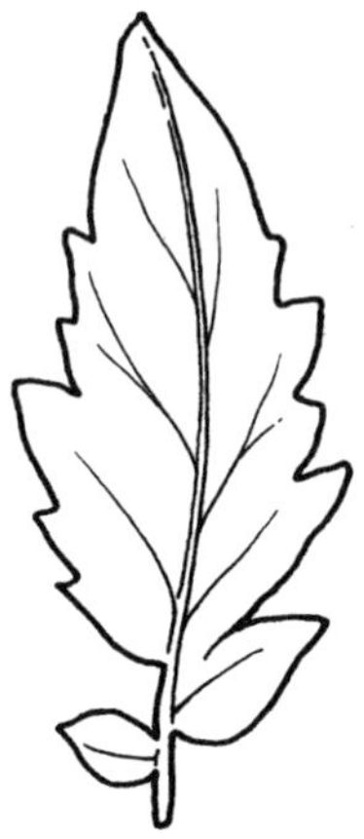

cut leaf

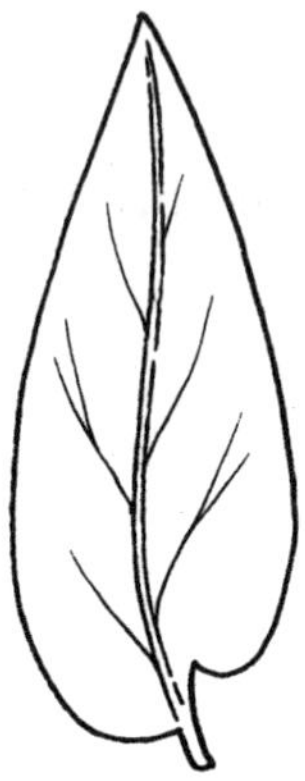

potato leaf

The parent plants were both true breeding.

The F_1 plants were self-pollinated.

The F_2 generation contained 80 plants with the cut leaf genotype and 20 plants with the potato leaf genotype.

(i) Explain what is meant by the term true breeding.

____ **1**

(ii) State which characteristic is dominant and give a reason.

Dominant characteristic __________________________

Reason ____ **1**

DO NOT WRITE IN THIS MARGIN

Marks

8. (*b*) (continued)

(iii) The results obtained in the F_2 generation differ from the expected results.

There were 100 plants in the F_2 generation.

Calculate the number of plants which would have been expected to have the cut leaf phenotype in the F_2 generation.

Space for calculation

__________ plants **1**

(iv) Suggest one reason why the results obtained differ from the expected results.

____ **1**

[Turn over

9. (a) The diagram below shows part of a food web from a moorland ecosystem.

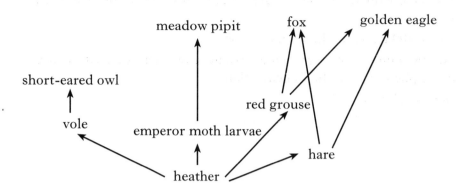

(i) Write each organism from the food web in the correct column in the table below.

Producer	Primary consumer	Secondary consumer

2

(ii) Describe **one** possible effect on the food web of a large increase in the grouse population.

_____ 1

(iii) Human activities can affect the biodiversity of an ecosystem.

Explain the term "biodiversity".

_____ 1

(b) Give one characteristic of a **stable** ecosystem.

_____ 1

10. The diagram below shows a genetically-engineered bacterial cell.

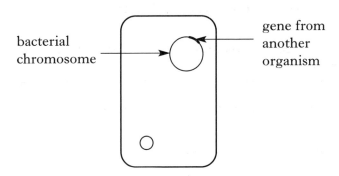

(a) Name the type of molecule found in a bacterial chromosome.

_____ **1**

(b) The bacterium was genetically engineered to produce a particular substance.

Name one example of a substance which can be produced in this way and describe its use.

Substance _____

Use _____

_____ **1**

(c) Genetic engineering can produce new genotypes to provide better organisms for particular functions.

Give one advantage and one disadvantage of genetic engineering compared with selective breeding.

Advantage _____

_____ **1**

Disadvantage _____

_____ **1**

[Turn over

11. (a) In 1994 a survey of habitats was carried out in West Lothian.

The table below shows the range of types of habitat in an area of 50 000 hectares.

The information was used to monitor the success of a biodiversity enhancement programme.

Type of habitat	Area (hectares)	Percentage cover
woodland and scrub	7 000	14·0
grassland	18 000	36·0
tall herb and fen	300	0·6
heathland	1 500	3·0
peatland	2 000	4·0
open water	500	
cultivated land	13 500	27·0
unsurveyed urban areas	6 600	13·2
others	600	1·2

(i) Complete the bar graph below to illustrate some of the information in the table.
(Additional graph paper, if required, will be found on page 28.)

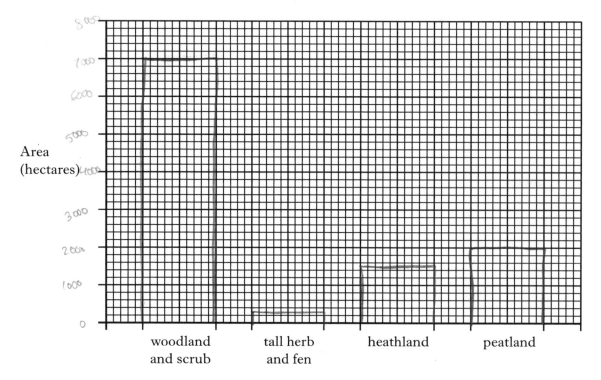

11. **(a)** **(continued)**

(ii) Calculate the percentage of the area which is open water.
Space for calculation

_____ % 1

(b) Name **one** human activity which could affect biodiversity in the open water habitat.

_____ 1

[Turn over

SECTION C

Both questions in this section should be attempted.

Note that each question contains a choice.

Questions 1 and 2 should be attempted on the blank pages which follow.

Supplementary sheets, if required, may be obtained from the invigilator.

Labelled diagrams may be included where appropriate.

In question 1, ONE mark is available for coherence.

1. Answer **either** A **or** B.

   A. The diagram below shows the appearance of human red blood cells in an isotonic solution.

   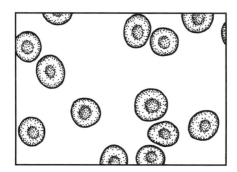

   Describe and explain the events which would take place if these red blood cells were transferred to pure water. 5

   OR

   B. The diagram below represents an amylase molecule.

   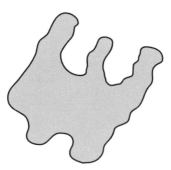

   Describe and explain the events that take place when the enzyme is added to a starch suspension. 5

Question 2 is on *page twenty-six*.

SPACE FOR ANSWER TO QUESTION 1

[Turn over for Question 2 on *page twenty-six*

2. Answer **either** A **or** B.

Labelled diagrams may be included where appropriate.

A. Describe the functions of macrophages and lymphocytes in defence. **5**

OR

B. Describe the functions of the hypothalamus and blood vessels in temperature regulation. **5**

[*END OF QUESTION PAPER*]

SPACE FOR ANSWER TO QUESTION 2

SPACE FOR ANSWERS

ADDITIONAL GRAPH PAPER FOR QUESTION 2(a)

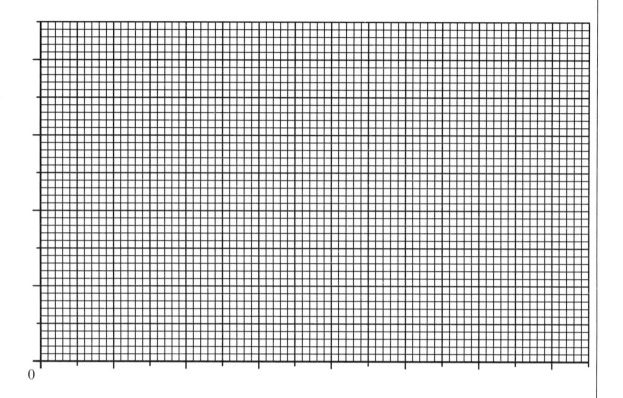

Temperature (°C)

ADDITIONAL GRAPH PAPER FOR QUESTION 11(a)(i)

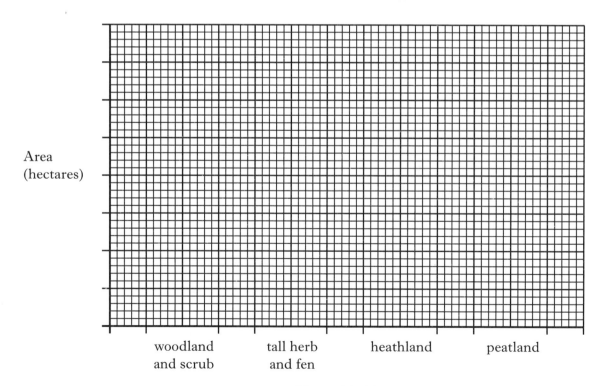

Area (hectares)

woodland and scrub   tall herb and fen   heathland   peatland

Type of habitat

2002 | Intermediate 2

FOR OFFICIAL USE

X007/201

Total for Sections B and C

NATIONAL QUALIFICATIONS 2002

FRIDAY, 31 MAY 1.00 PM – 3.00 PM

BIOLOGY INTERMEDIATE 2

Fill in these boxes and read what is printed below.

Full name of centre

Town

Forename(s)

Surname

Date of birth
Day Month Year   Scottish candidate number   Number of seat

SECTION A (25 marks)
Instructions for completion of Section A are given on page two.

SECTIONS B AND C (75 marks)

1. (a) All questions should be attempted.

   (b) It should be noted that in **Section C** questions 1 and 2 each contain a choice.

2. The questions may be answered in any order but all answers are to be written in the spaces provided in this answer book, and must be written clearly and legibly in ink.

3. Additional space for answers and rough work will be found at the end of the book. If further space is required, supplementary sheets may be obtained from the invigilator and should be inserted inside the **front** cover of this book.

4. The numbers of questions must be clearly inserted with any answers written in the additional space.

5. Rough work, if any should be necessary, should be written in this book and then scored through when the fair copy has been written.

6. Before leaving the examination room you must give this book to the invigilator. If you do not, you may lose all the marks for this paper.

Read carefully

1 Check that the answer sheet provided is for Biology Intermediate 2 (Section A).

2 Fill in the details required on the answer sheet.

3 In this section a question is answered by indicating the choice A, B, C or D by a stroke made in **ink** in the appropriate place in the answer sheet—see the sample question below.

4 For each question there is only **one** correct answer.

5 Rough working, if required, should be done only on this question paper, or on the rough working sheet provided—**not** on the answer sheet.

6 At the end of the examination the answer sheet for Section A **must** be placed inside the front cover of this answer book.

Sample Question

Which of the following lists all the elements that are present in every protein molecule?

A Carbon, oxygen, nitrogen

B Carbon, hydrogen, oxygen, nitrogen

C Carbon, hydrogen, oxygen, sulphur

D Carbon, hydrogen, oxygen

The correct answer is B—Carbon, hydrogen, oxygen, nitrogen. A **heavy** vertical line should be drawn joining the two dots in the appropriate box in the column headed **B** as shown **in the example on the answer sheet**.

If, after you have recorded your answer, you decide that you have made an error and wish to make a change, you should cancel the original answer and put a vertical stroke in the box you now consider to be correct. Thus, if you want to change an answer **D** to an answer **B**, your answer sheet would look like this:

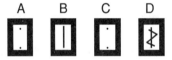

If you want to change back to an answer which has already been scored out, you should **enter a tick (✓)** to the RIGHT of the box of your choice, thus:

SECTION A

All questions in this Section should be attempted.

Questions 1 and 2 refer to the diagram of the digestive system below.

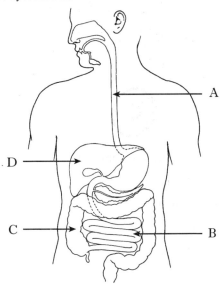

1. Which labelled arrow identifies the small intestine?

2. Which labelled arrow identifies where water is absorbed?

3. The diagram below shows some structures in a villus.

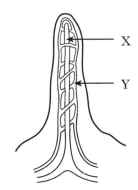

Which line in the table below correctly identifies the products of digestion which pass into structures X and Y?

	X	Y
A	glucose	amino acids
B	glycerol	fatty acids
C	amino acids	glycogen
D	fatty acids	glucose

4. The key below can be used to identify four components of blood, P, Q, R and S.

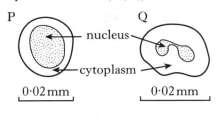

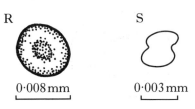

1. Nucleus present............................. go to 2
   Nucleus absent.............................. go to 3

2. Large volume of cytoplasm present .. macrophage
   Small volume of cytoplasm present .. lymphocyte

3. Diameter greater than 0·005 mm ...... red blood cell
   Diameter less than 0·005 mm .... platelet

Which line in the table correctly identifies the blood components?

	P	Q	R	S
A	lymphocyte	red blood cell	platelet	macrophage
B	macrophage	lymphocyte	red blood cell	platelet
C	platelet	macrophage	red blood cell	lymphocyte
D	lymphocyte	macrophage	red blood cell	platelet

[Turn over

5. The diagrams below show stages in the process of phagocytosis as carried out by a macrophage.

1. The bacterium is trapped in the vacuole.

2. A vacuole forms in the macrophage.

3. The bacterium is digested by enzymes.

4. The macrophage engulfs the bacterium.

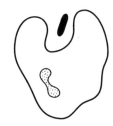

The correct order for these stages is

A 4 1 3 2

B 2 4 1 3

C 4 2 1 3

D 1 3 2 4.

Questions 6 and 7 refer to the diagrams of cells below.

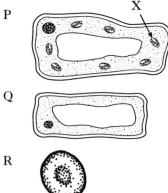

6. Which cells are plant cells?

A P only

B P and Q

C P and R

D R only

7. The function of structure X is to

A control all cell activities

B keep the cell turgid

C produce glucose using light energy

D release energy from glucose.

8. Lipase is an enzyme found in the small intestine. Lipase speeds up the breakdown of fat. Full cream milk contains a high proportion of fat.

Three test tubes were set up as shown in the diagrams below.

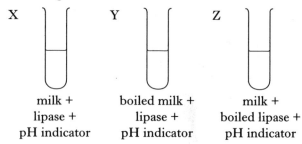

X	Y	Z
milk + lipase + pH indicator	boiled milk + lipase + pH indicator	milk + boiled lipase + pH indicator

The pH of the contents of each test tube was recorded at the start and again 15 minutes later.

What changes in pH took place?

A The pH decreased in each test tube.

B The pH increased in each test tube.

C The pH decreased in tubes X and Y and did not change in tube Z.

D The pH increased in tubes Y and Z and did not change in tube X.

Questions 9 and 10 refer to the bar chart below. The bar chart shows the volume of blood supplied per minute to the skeletal muscles and to other parts of the body of a healthy male at rest and during strenuous exercise.

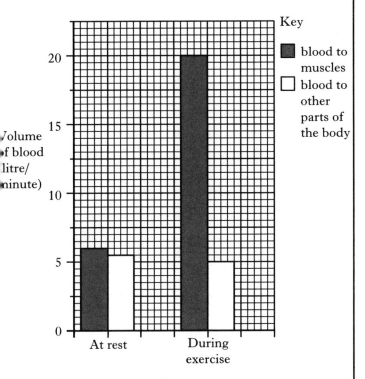

9. There is a difference in the volume of blood supplied per minute to the muscles and to other parts of the body **at rest**.

   Which statement below is correct?

   A 0·5 litre more blood is supplied to the muscles than to other parts of the body.

   B 1·0 litre less blood is supplied to the muscles than to other parts of the body.

   C 0·5 litre less blood is supplied to the muscles than to other parts of the body.

   D 1·0 litre more blood is supplied to the muscles than to other parts of the body.

10. During **exercise**, the ratio of blood supplied to the muscles to blood supplied to other parts of the body is

    A 1 : 4
    B 4 : 1
    C 6 : 5
    D 10 : 3.

11. Haemoglobin combines with oxygen to form oxy-haemoglobin.

    Which of the following statements is correct?

    A At low oxygen levels oxy-haemoglobin releases oxygen in the lungs.

    B At high oxygen levels oxy-haemoglobin releases oxygen in the tissues.

    C At low oxygen levels oxy-haemoglobin releases oxygen in the tissues.

    D At high oxygen levels oxy-haemoglobin releases oxygen in the lungs.

Questions 12 and 13 refer to the diagram of the brain below.

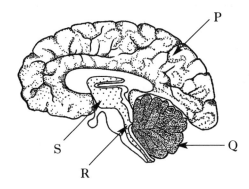

12. The part labelled **P** is the

    A cerebrum
    B cerebellum
    C medulla
    D hypothalamus.

13. The hypothalamus

    A is responsible for conscious responses

    B co-ordinates activities in all parts of the body

    C detects changes in the water content of the blood

    D controls the heart and breathing rates.

[Turn over

14. Dried yeast was mixed with flour and sugar solution to make dough.

The dough was put into a measuring cylinder as shown in the diagram below.

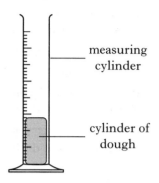

The volume of dough was measured over a 40 minute period.

The results are shown in the table below.

Time (minutes)	0	10	20	30	40
Volume of dough (cm³)	25	27	31	37	40

The percentage increase in the volume of the dough from the start to the end of the period was

A 15

B 37·5

C 60

D 62·5.

15. Which of the following statements is correct?

A Anaerobic respiration produces 38 molecules of ATP from each glucose molecule.

B Anaerobic respiration produces twice as much energy as aerobic respiration.

C Aerobic respiration produces 38 molecules of ATP from each glucose molecule.

D Aerobic respiration produces 2 molecules of ATP from each glucose molecule.

16. Which of the following is a **reversible** reaction in anaerobic respiration?

A The conversion of pyruvic acid to ethanol and carbon dioxide

B The conversion of glucose to pyruvic acid

C The conversion of pyruvic acid to carbon dioxide and water

D The conversion of pyruvic acid to lactic acid

17. A piece of potato was cut from a potato tuber, blotted dry and weighed.

It was placed in pure water for an hour and then removed, dried and re-weighed.

It was then placed in concentrated sugar solution for an hour, removed, dried and re-weighed.

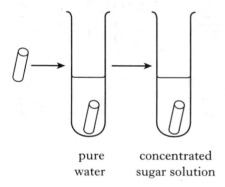

Which line in the table shows the results most likely to be obtained?

	First weight (g)	Second weight (g)	Third weight (g)
A	5	6	4
B	5	4	6
C	5	7	9
D	5	4	3

Questions 18 and 19 refer to the diagram below.

The diagram represents energy flow in a woodland ecosystem.

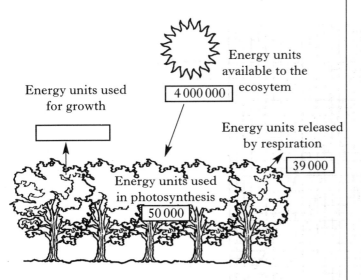

18. The number of energy units used for growth is

    A   11 000
    B   89 000
    C   3 950 000
    D   3 961 000.

19. The percentage of the energy from sunlight absorbed by trees and used for photosynthesis is

    A   1·25
    B   12·5
    C   98·75
    D   8000.

20. The diagram below shows the pyramid of energy for a food chain.

    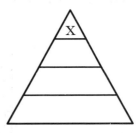

    There is less energy at level X in the pyramid because

    A   energy is stored in each level and not passed on
    B   energy is lost at each level in a food chain
    C   the energy is concentrated in fewer organisms
    D   organisms in level X are very small.

21. The table below shows the relationship between planting density and the mass of seed harvested for a trial cereal crop.

Planting density (number of plants per square metre)	Mass of seed harvested (grammes per square metre)
4	60
8	86
16	104
32	77
128	22

    The reason a low mass of seed was harvested when the planting density was 128 plants per square metre was

    A   less disease at high planting densities
    B   more nutrients available
    C   more competition for light and nutrients
    D   less space for weeds.

[Turn over

Questions **22** and **23** refer to the graph below.

The graph shows the variations in the populations of arctic hares and their predators, foxes, in an area.

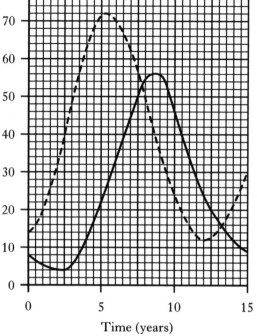

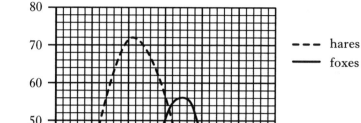

22. What is the difference in the population sizes of the hares and foxes when both populations are at their maximum?

A 3 000

B 16 000

C 52 000

D 58 000

23. What could be the reason for the decline in number of foxes after 9 years?

A The food available decreased.

B The numbers of prey increased.

C There is less disease in both populations.

D The food available increased.

24. The table below gives information about chromosomes in some human cells.

Which line in the table is correct?

	Cell	Number of chromosome sets	Number of chromosomes
A	nerve	2	23
B	egg	1	23
C	lymphocyte	1	46
D	sperm	2	23

25. When a plant with red flowers was crossed with a plant with white flowers, the F_1 plants had pink flowers.

The F_1 plants were then self-fertilised.

What ratio of flower colours would be found in the F_2 generation?

A 1 red : 2 white

B 1 white : 2 pink

C 1 red : 2 pink : 1 white

D 1 red : 1 pink : 1 white

Candidates are reminded that the answer sheet for Section A MUST be placed inside the front cover of this answer book.

SECTION B

All questions in this section should be attempted.

1. (a) The diagram below shows a capillary network similar to that found surrounding air sacs in the lungs.

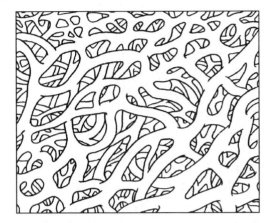

State **two** features of the **capillary network** in the lungs that allow efficient gas exchange to take place.

Explain how each feature improves gas exchange.

Feature 1 _____ 1

Explanation _____ 1

Feature 2 _____ 1

Explanation _____ 1

(b) The diagram below shows gas exchange taking place in the tissues.

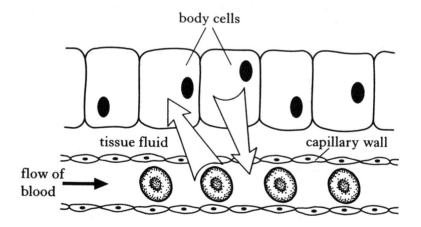

Name the process by which gases pass between capillaries and body cells.

_____ 1

2. The bar graph below shows average blood pressure measurements in different blood vessels for a healthy individual.

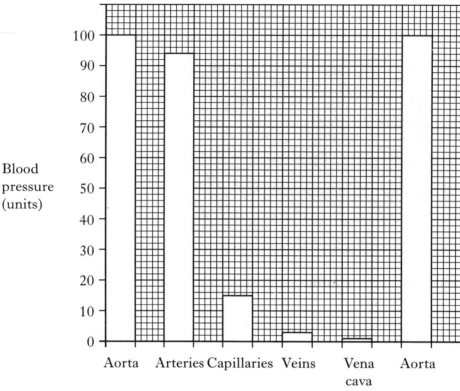

Site of blood pressure measurement

(a) Calculate the **simple whole number ratio** of the blood pressure in the aorta to the blood pressure found in the capillaries.

Space for calculation

_____ : _____

(b) Calculate the **percentage decrease** in blood pressure when blood moves from a capillary to a vein.

Space for calculation

_____ %

(c) Explain what brings about the large increase in blood pressure between the vena cava and the aorta.

_____

(d) The blood pressure in veins is very low.

Name the structures in veins that prevent blood flowing backwards.

_____

3. An investigation was carried out into the concentration of lactic acid in the blood before, during and after a two minute period of strenuous exercise.

The results are shown in the graph below.

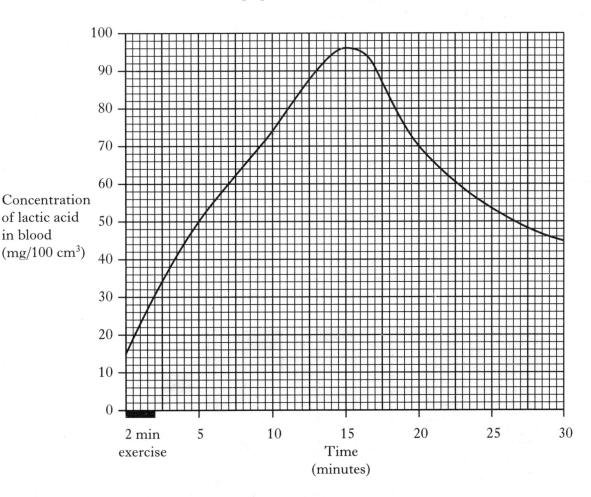

(a) What was the concentration of lactic acid in the blood ten minutes after the start of the exercise?

_____ mg/100 cm³

(b) Explain why the concentration of lactic acid in the blood increased during and immediately after the strenuous exercise.

_____

_____

(c) What caused the lactic acid concentration to decrease 15 minutes after the start of the exercise?

_____

_____

4. (a) The table below shows the composition of four foods.

Food	Food Components				
	Carbohydrate %	Fat %	Protein %	Water %	Other %
Meat	0	18	18	62	2
Milk	5	5	4	86	0
Maize	10	2	7	78	3
Soya Beans	34	18	34	9	5

(i) Complete the bar graph by

1. putting a scale on the vertical axis
2. adding the bar to show all the components of meat.

(Additional graph paper, if required, will be found on page 32.)

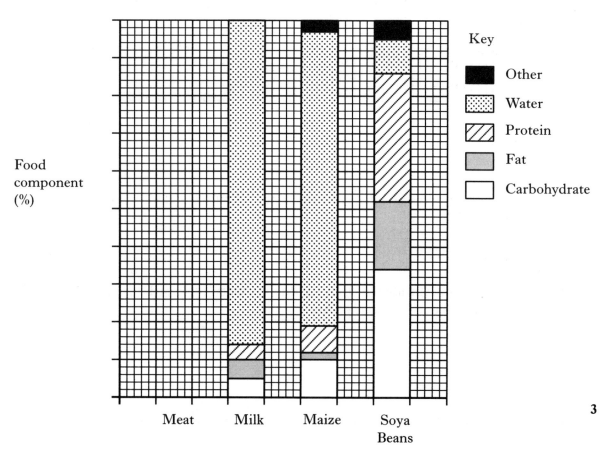

DO NOT WRITE IN THIS MARGIN

Marks

4. (*a*) (continued)

(ii) The final column in the table has been headed "other".

Name **one** of the food groups which is included under this heading.

______________________________ **1**

(iii) Which component of food is needed to produce enzymes?

______________________________ **1**

(*b*) Name the **three** elements common to carbohydrate, protein and fat.

____ **1**

[Turn over

5. The diagram below shows some of the structures involved in a reflex action.

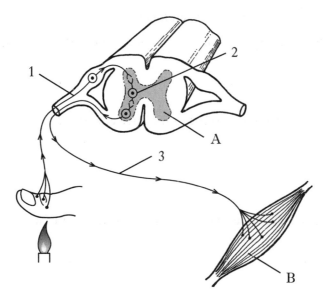

(a) The neurones labelled 1, 2 and 3 form a reflex arc.

Name each of these neurones.

1 _____

2 _____

3 _____   2

(b) Some neurones found in area A may transmit impulses to another part of the Central Nervous System (CNS).

Name the part of the CNS which receives these impulses.

_____   1

(c) Describe the response that occurs at B.

_____   1

(d) What is the function of reflex actions?

_____   1

6. (a) A freshwater fish is hypertonic to its environment.

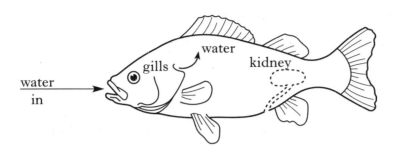

Underline **one** option in each set of brackets to make the sentences below correct.

The gills of this fish { take in / give out } salts.

The kidneys produce a { large / small } volume of { dilute / concentrated } urine.

(b) (i) Name the hormone that acts on the kidney to maintain water balance in humans.

_____

(ii) Name the part of the kidney nephron on which the hormone has its effect.

_____

(iii) When the concentration of water in the blood increases, the hormone acts to return the water concentration to normal.

What term is used to describe this mechanism?

_____

[Turn over

7. Albumin is a protein which can be broken down by the enzyme trypsin.

When albumin is added to agar the agar becomes cloudy.

When the albumin is broken down by trypsin the agar becomes clear.

$$albumin \xrightarrow{trypsin} amino\ acids$$

Groups of students carried out an investigation to find the effect of pH on the activity of the enzyme trypsin.

Each group carried out the following procedure.

1. A petri dish containing cloudy albumin agar was collected.
2. Three wells were cut out of the agar.
3. Drops of trypsin and buffer solution of different pH were added to the wells, as shown in the diagram below.

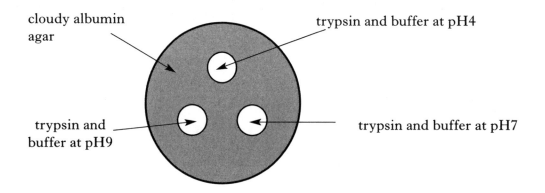

4. The dishes were incubated at 37 °C for 24 hours.
5. The diameter of the clear zone around each well was measured.

(a) State **one** variable that should have been kept constant when the dishes were set up.

_____ 1

7. (continued)

(b) The results from each group are given in the table below.

	Diameter of the clear area (mm)				
pH	Group 1	Group 2	Group 3	Group 4	Average
4	1	2	0	3	1·5
7	6	9	8	11	
9	21	23	20	24	22

(i) Complete the table by calculating the average diameter of the clear area for pH7.

Space for calculation

(ii) Explain why the results for all groups were collected and averages calculated.

_____

(iii) From the results, describe the effect of pH on the activity of trypsin.

_____

(c) One group suggested setting up a control to show that the enzyme caused the observed effects.

Complete the labels on the diagram below to show the content of the wells for the control.

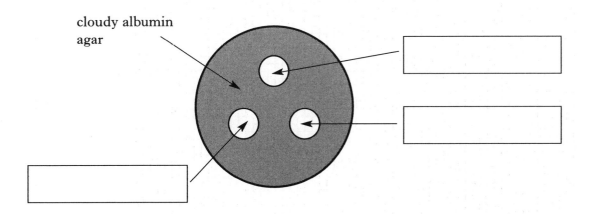

8. (a) A scientist grew some cereal plants in a field.

During the course of a day, she removed 2 plants every 4 hours and measured the concentrations of sugar in the leaves of the plants.

The results are shown in the table below.

Time of day (hours)	Sugar concentration (percentage of dry mass)		
	Sample 1	Sample 2	Average
0400	0·42	0·48	0·45
0800	0·58	0·62	0·60
1200	1·46	2·04	1·75
1600	1·57	2·43	2·00
2000	1·05	1·75	1·40
2400	0·49	0·51	0·50

Present the results in an appropriate format on the grid below.
(Additional graph paper, if required, will be found on page 33.)

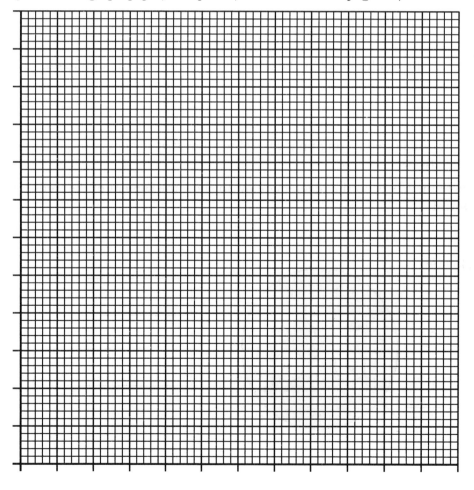

DO NOT WRITE IN THIS MARGIN

Marks

8. (continued)

(*b*) The following word equation shows the first stage of photosynthesis.

$$\text{water} \xrightarrow{\text{light energy}} \text{oxygen} + \text{hydrogen} + \text{ATP}$$

(i) Name this stage of photosynthesis.

____________________________ **1**

(ii) Describe what happens to each of the products.

Oxygen ____

____ **1**

Hydrogen ____

____ **1**

ATP ____

____ **1**

[Turn over

9. (a) Lettuces may be grown in "tunnels" covered in clear polythene as shown in the diagram below.

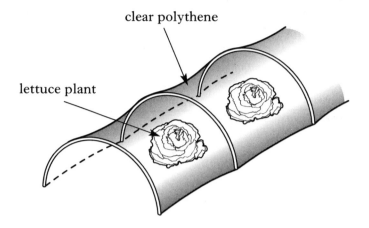

(i) Give **two** reasons why the use of clear polythene may result in the lettuces being ready for cropping earlier.

1 _____

2 _____   2

(ii) Air rich in carbon dioxide can be passed through these tunnels.

Explain how this would make the lettuces grow faster.

_____   1

9. (continued)

(b) The graph below shows how the rate of photosynthesis is affected by light intensity at different concentrations of carbon dioxide.

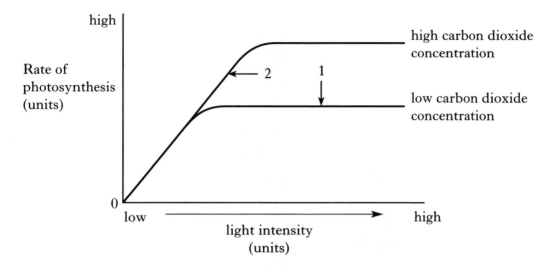

(i) What factor was limiting the rate of photosynthesis at points 1 and 2?

Point 1 _____ 1

Point 2 _____ 1

(ii) Name **one** other factor that may limit the rate of photosynthesis.

_____ 1

[Turn over

10. (*a*) In peas the height of the plant is controlled by one gene which has two alleles.

T represents the dominant allele for tall stems.

t represents the allele for short stems.

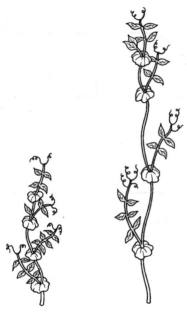

True breeding, tall-stemmed pea plants were crossed with short-stemmed pea plants to produce the F_1 generation.

(i) State the genotype of the parents.

_____ and _____

(ii) State the phenotypes of the F_1 plants.

_____

(iii) Plants from the F_1 generation were crossed to produce the F_2 generation of plants.

State the phenotypes and their expected ratio in the F_2 generation.

_____ : _____ plants

10. (continued)

(b) Seeds from true breeding, tall-stemmed pea plants were provided with different growing conditions as shown in the diagrams below.

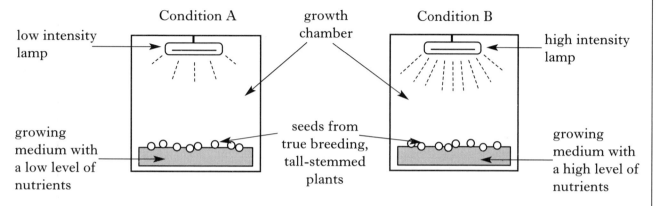

(i) **Compare** the expected appearance of the plants in each group.

_____

_____ 1

(ii) What term is used to describe the effect of different conditions on the phenotype of an organism?

_____ 1

[Turn over

Marks

11. The song thrush feeds on snails which it finds by sight in a variety of habitats. Snails are herbivores and feed only on grasses.

(a) (i) Present this feeding relationship as a food chain.

1

(ii) Use the information above to complete the table below.

Term	Organism
primary consumer	snail
predator	
prey	
producer	

2

(iii) Describe what is meant by a *habitat*.

_____ 1

(b) Name a decomposer and describe the role of decomposers in an ecosystem.

Decomposer _____ 1

Role of decomposers _____ 1

11. (continued)

(c) The snail *Cepea* shows variation in shell colour and different forms of shell are found.

Two of the forms are shown below.

Light form   Dark form

The table below shows the results of an investigation into the numbers of different forms of snails found in two habitats.

Habitat	Light form	Dark form	Other forms
dense woodland	14	64	22
open grassland	58	12	30

Explain how natural selection has resulted in the high numbers of the dark form of the shell in the **dense woodland**.

2

[Turn over

12. An investigation into the behaviour of blowfly larvae was carried out.

One blowfly larva was placed at X and then lamp A was switched on.

When the larva reached Y, lamp A was switched off and lamp B was switched on.

The path taken by the larva is shown in the diagram below.

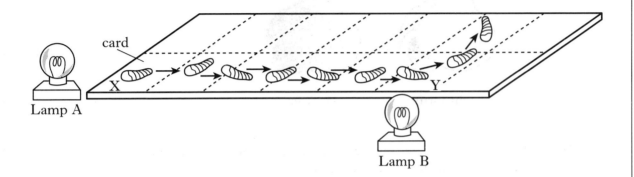

(a) Describe the response of the larva to light.

_____ 1

(b) (i) Suggest **one** change to the apparatus that would confirm that the response was due to the light and not the heat from the lamp.

_____ 1

(ii) State **one** way in which the reliability of the result could be improved.

_____ 1

(c) (i) Name a stimulus, other than light, to which woodlice respond.

_____ 1

(ii) Describe the response of woodlice to this stimulus.

_____ 1

(iii) Explain the adaptive significance of this response in woodlice.

_____ 1

[Turn over for SECTION C on *Page twenty-eight*

SECTION C

Both questions in this section should be attempted.

Note that each question contains a choice.

Questions 1 and 2 should be attempted on the blank pages which follow.

Supplementary sheets, if required, may be obtained from the invigilator.

In question 1, ONE mark is available for coherence.

1. Answer **either** A **or** B.

   A. The maps below show the changes in distribution of the red squirrel in the UK from 1920 to 1990.

   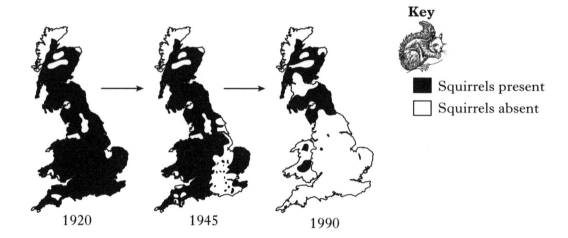

   (a) Describe the change in distribution between 1920 and 1990.

   (b) Suggest reasons for this change.

   (c) Discuss the potential impact this could have on biodiversity. **5**

OR

   B. The diagram below shows the two stages that result in the formation of offspring in animals.

   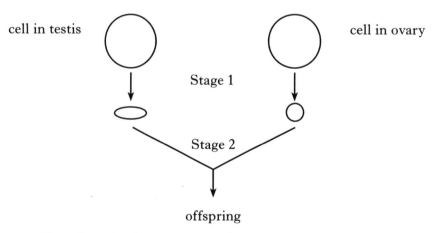

   (a) Describe what happens at **each** stage.

   (b) Explain how variation is brought about during **each** stage. **5**

Question 2 is on *Page thirty*.

SPACE FOR ANSWER TO QUESTION 1

2. Answer **either** A **or** B.

Labelled diagrams may be included where appropriate.

A. Give an account of the properties of enzymes. 5

OR

B. Describe the stages in aerobic respiration. 5

[END OF QUESTION PAPER]

SPACE FOR ANSWER TO QUESTION 2

SPACE FOR ANSWERS

ADDITIONAL GRAPH PAPER FOR QUESTION 4(a)(i)

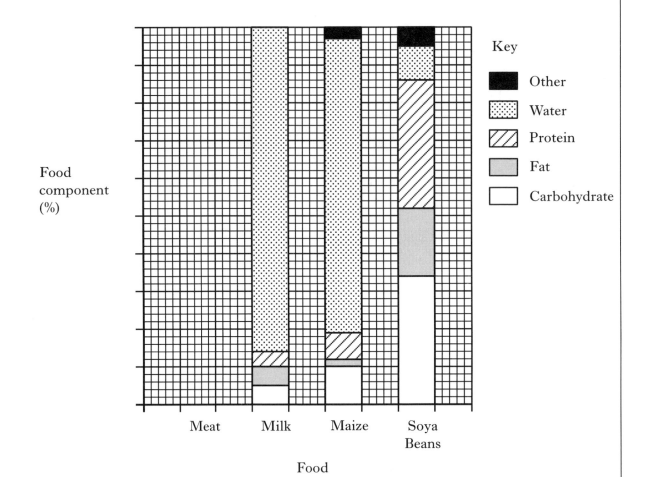

SPACE FOR ANSWERS

ADDITIONAL GRAPH PAPER FOR QUESTION 8(a)

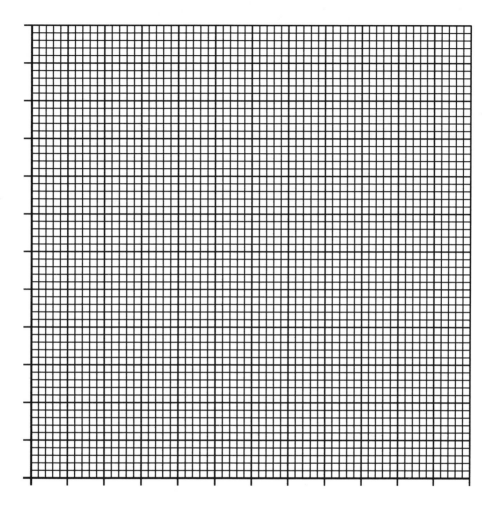

SPACE FOR ANSWERS

2003 | Intermediate 2

[BLANK PAGE]

FOR OFFICIAL USE

X007/201

Total for Sections B and C

NATIONAL
QUALIFICATIONS
2003

MONDAY, 26 MAY
9.00 AM – 11.00 AM

BIOLOGY
INTERMEDIATE 2

Fill in these boxes and read what is printed below.

Full name of centre

Town

Forename(s)

Surname

Date of birth
Day Month Year

Scottish candidate number

Number of seat

SECTION A (25 marks)

Instructions for completion of Section A are given on page two.

SECTIONS B AND C (75 marks)

1 (a) All questions should be attempted.

   (b) It should be noted that in **Section C** questions 1 and 2 each contain a choice.

2 The questions may be answered in any order but all answers are to be written in the spaces provided in this answer book, and must be written clearly and legibly in ink.

3 Additional space for answers and rough work will be found at the end of the book. If further space is required, supplementary sheets may be obtained from the invigilator and should be inserted inside the **front** cover of this book.

4 The numbers of questions must be clearly inserted with any answers written in the additional space.

5 Rough work, if any should be necessary, should be written in this book and then scored through when the fair copy has been written.

6 Before leaving the examination room you must give this book to the invigilator. If you do not, you may lose all the marks for this paper.

Read carefully

1. Check that the answer sheet provided is for Biology Intermediate 2 (Section A).

2. Fill in the details required on the answer sheet.

3. In this section a question is answered by indicating the choice A, B, C or D by a stroke made in **ink** in the appropriate place in the answer sheet—see the sample question below.

4. For each question there is only **one** correct answer.

5. Rough working, if required, should be done only on this question paper, or on the rough working sheet provided—**not** on the answer sheet.

6. At the end of the examination the answer sheet for Section A **must** be placed inside the front cover of this answer book.

Sample Question

What must be present in leaf cells for photosynthesis to take place?

A Oxygen and water

B Carbon dioxide and water

C Carbon dioxide and oxygen

D Oxygen and hydrogen

The correct answer is B—Carbon dioxide and water. A **heavy** vertical line should be drawn joining the two dots in the appropriate box in the column headed **B** as shown **in the example on the answer sheet**.

If, after you have recorded your answer, you decide that you have made an error and wish to make a change, you should cancel the original answer and put a vertical stroke in the box you now consider to be correct. Thus, if you want to change an answer **D** to an answer **B**, your answer sheet would look like this:

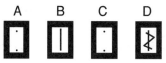

If you want to change back to an answer which has already been scored out, you should **enter a tick (✓)** to the RIGHT of the box of your choice, thus:

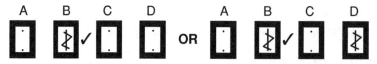

SECTION A

All questions in this Section should be attempted.

1. Which carbohydrate is a component of cell walls?
   A Glycogen
   B Starch
   C Cellulose
   D Glucose

2. Enzymes act as catalysts because they
   A are composed of protein
   B act on all substrates
   C raise energy input
   D lower energy input.

3. The active site of an enzyme is complementary to
   A one type of substrate molecule
   B all types of substrate molecules
   C one type of product molecule
   D all types of product molecules.

4. Four thin sections of onion tissue were immersed in 5% sugar solution. The sections were left for 15 minutes then viewed under the microscope. The table shows the percentage of cells plasmolysed in each section.

Section	Cells plasmolysed (%)
1	22
2	22
3	27
4	29

   The average percentage of cells plasmolysed is
   A 22
   B 25
   C 27
   D 100.

5. The breakdown of ATP in cells
   A releases energy and produces ADP only
   B releases energy and produces ADP + P_i
   C requires energy and produces ADP only
   D requires energy and produces ADP + P_i.

6. How many <u>more</u> ATP molecules are produced per glucose molecule by aerobic respiration than by anaerobic respiration?
   A 2
   B 19
   C 36
   D 38

7. Which of the following conditions in a greenhouse would produce earlier crops?
   A Glass shading
   B Cool air conditioners
   C Additional oxygen
   D Additional carbon dioxide

8. The diagram below shows a virus attached to a blood cell. The blood cell has responded by producing specific protein molecules labelled X.
   (Diagram not drawn to scale.)

   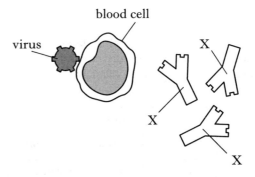

   The molecules labelled X are
   A antibodies
   B antigens
   C lymphocytes
   D macrophages.

9. The table below refers to information about a breakfast cereal.

Ingredients	Mass per serving
Protein	6 g
Carbohydrate	62 g
Fat	4 g
Vitamins	1·4 mg
Iron	2·4 mg

One serving will provide 20% of a child's daily requirement for iron.

How many mg of iron are required daily by a child?

A   0·12
B   0·48
C   12
D   48

10. The diagram below shows the movement of food along the oesophagus.

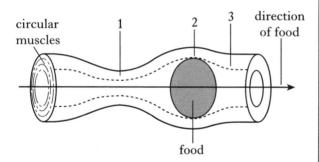

Which line in the table below correctly describes the state of the circular muscles at points 1, 2 and 3 on the diagram?

	Circular muscles		
	Point 1	Point 2	Point 3
A	contracted	relaxed	contracted
B	relaxed	contracted	contracted
C	contracted	relaxed	relaxed
D	relaxed	contracted	relaxed

11. The following graph shows the results of an investigation into the effect of pH on the activity of four enzymes.

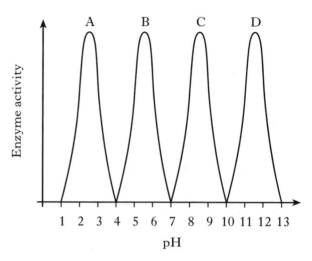

Which one of these enzymes could be pepsin in the stomach?

12. Which label correctly identifies the lacteal in the following diagram of a villus?

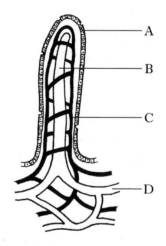

13. Which line in the table below correctly describes what happens to excess proteins in the diet?

	Site of deamination	Product
A	liver	urea
B	kidney	urea
C	liver	amino acids
D	kidney	amino acids

14. A food contains the elements carbon, hydrogen, oxygen and nitrogen. To which food group does it belong?

A Carbohydrates

B Proteins

C Fats

D Minerals

Questions 15 and 16 refer to the table below which shows the composition of the blood entering the kidney and the composition of the urine leaving the kidney.

Substances	Composition of blood entering the kidney (%)	Composition of urine leaving the kidney (%)
Water	92	95
Protein	7	0
Glucose	0·10	0
Salts	0·37	0·60
Urea	0·03	2·00

15. Which of the following substances are all excreted by the kidney?

A Water, glucose and salts

B Water, salts and urea

C Salts, protein and urea

D Salts, glucose and protein

16. How many times greater is the urea concentration in urine than in blood?

A 0·015

B 0·06

C 1·97

D 66·67

Questions 17 and 18 refer to the graph below which shows changes in blood pressure in the aorta during one heart beat cycle.

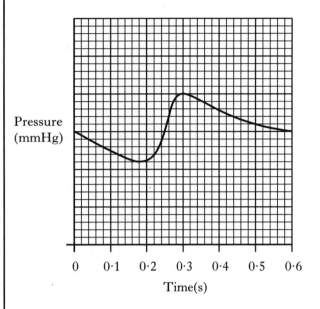

17. What is the heart rate in beats per minute?

A 30

B 60

C 100

D 120

18. At what time do the ventricles start to contract?

A 0·1 s

B 0·2 s

C 0·3 s

D 0·4 s

[Turn over

19. The diagram below shows a human sperm, egg and female zygote.

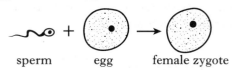

Which line in the table correctly describes the sex chromosomes in each of these cells?

	Sex chromosome(s) of sperm	Sex chromosome(s) of egg	Sex chromosome(s) of female zygote
A	Y	X	XY
B	XY	XX	Y
C	XX	XY	X
D	X	X	XX

20. A species can be defined as a group of organisms which

A contain identical genetic material

B have the same phenotypes

C contain the same number of chromosomes

D breed together to produce fertile offspring.

Questions 21 and 22 refer to the following statements about a woodland ecosystem.

A All the oak trees

B All the plants

C All the plants and animals

D All the oak trees and blackbirds

21. Which statement describes a population?

22. Which statement describes a community?

23. A sample of fresh soil from a woodland ecosystem was weighed, dried in an oven at 95 °C for one week and reweighed.
The results are shown below.

Mass of fresh soil = 50 g
Mass of dried soil = 32 g

What percentage of the soil sample was water?

A 9

B 18

C 36

D 64

24. Which one of the following graphs shows the effects of competition for the same food between a successful species and an unsuccessful species?

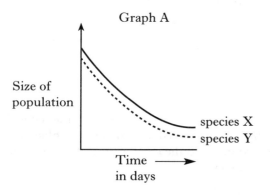

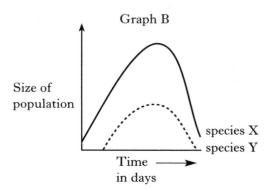

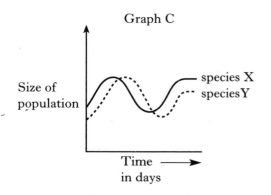

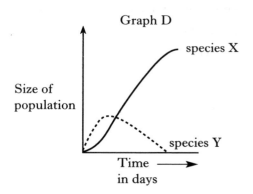

25. A river was sampled at six points along its length. The numbers of different animals, the oxygen concentration and the pH were recorded for each sampling point.

The results are shown in the table below.

	Sampling points					
	1	2	3	4	5	6
Mayfly nymphs	0	0	0	5	6	132
Dragonfly nymphs	1	1	0	0	1	1
Chironimid fly larvae	0	1	1	2	231	36
Molluscs	0	0	0	0	46	73
Oxygen concentration (%)	88	80	75	71	30	63
pH	5·6	6·0	6·5	7·3	7·5	8·0

Using these results identify which of the following conclusions is **correct**.

A Chironimid fly larvae do not survive in water of a low oxygen concentration.

B Molluscs survive better in water of a lower pH.

C The distribution of Dragonfly nymphs is not affected by changes in the pH and oxygen concentration of the water.

D The distribution of Mayfly nymphs is not affected by the oxygen concentration of the water.

Candidates are reminded that the answer sheet for Section A MUST be placed <u>inside</u> the front cover of this answer book.

[Turn over for Section B on *Page nine*

[BLANK PAGE]

SECTION B

All questions in this section should be attempted.

1. The diagram below shows a section through a plant cell.

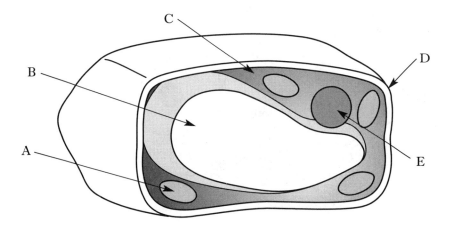

   (a) (i) Which **two** letters identify structures found in both plant and animal cells?

   _____

   _____

   (ii) Name the enzyme-controlled process associated with structure A.

   _____

   (b) Name a molecule found in structure E which is composed of a sequence of bases.

   _____

[Turn over

2. The diagram below represents a section of human tissue showing an exchange of materials between the body cells and blood.

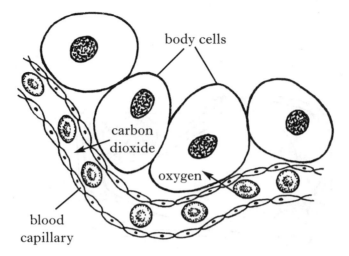

(a) Name and describe the process by which carbon dioxide moves out of the body cells into the blood.

Name of process _____

Description of process _____

_____

(b) Why is it important that carbon dioxide is removed from the body cells?

_____

_____

(c) Name the cell process which uses oxygen as a raw material.

_____

3. Three discs were cut from the same potato and were placed in three salt solutions of different concentrations. After 30 minutes the discs were removed from the solutions and the cells examined under a light microscope. A cell from each disc is shown below.

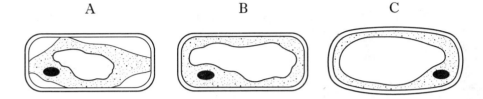

(a) (i) Identify the cell which was placed in

a hypertonic solution ———

an isotonic solution ———

(ii) Name the process which causes the difference in appearance of the cells.

————————————————

(iii) What name is used to describe the condition of cell C?

————————————————

(b) Name the cell structure which prevents plant cells from bursting.

————————————————

(c) Describe the appearance of red blood cells when placed in a hypertonic solution.

————————————————

(d) Name the enzyme which catalyses the synthesis of starch in potato cells.

————————————————

[Turn over

4. (a) The corncrake is a bird once found throughout the UK, but now mostly found in the north and west of Scotland.

The decrease in corncrake numbers was caused by a change in hay cutting methods.

Different farming methods were introduced from 1992 to save the corncrake.

The following table shows the estimated numbers of adult corncrake males in Scotland from 1988 to 2001.

Year	Estimated number of adult males
1988	540
1990	485
1992	440
1994	470
1996	510
1999	590
2001	600

(i) Present the results in an appropriate format on the grid below. (Additional graph paper, if required, will be found on page 32.)

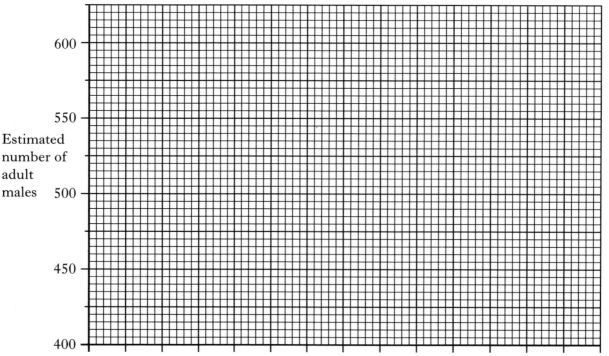

4. *(a)* **(continued)**

(ii) Describe the effect of the introduction of different farming methods on the corncrake population.

_____

_____ 1

(b) The change in the corncrake population is the result of human activity. This affects biodiversity.

Give **one** other example of a human activity that affects biodiversity and describe the effect.

Human activity _____ 1

Effect on biodiversity _____

_____ 1

(c) The bar chart below illustrates biodiversity in three different meadows.

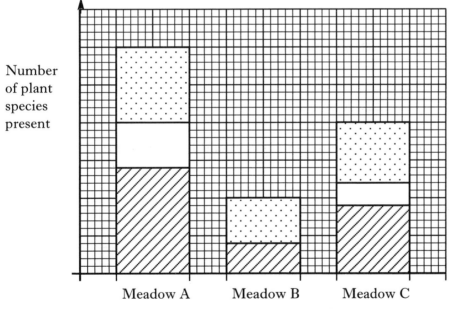

Which meadow has the lowest intensity of grazing?

_____ 1

(d) Describe an adaptation of a desert plant and explain how this adaptation aids survival in desert conditions.

Adaptation _____ 1

Explanation _____

_____ 1

5. Brine shrimps are invertebrates that live in salt water. They feed on microscopic green plants filtered from the water.

An investigation into the effect of light on the behaviour of brine shrimps was carried out by five groups of students. The following description and diagram detail how this investigation was set up by each group.

1. A petri dish was half-filled with salt water and six brine shrimps were added.
2. The brine shrimps were allowed to swim around for two minutes.
3. Half of the petri dish was covered in black paper.
4. After a further two minutes the number of brine shrimps found in the light and dark sides was recorded.

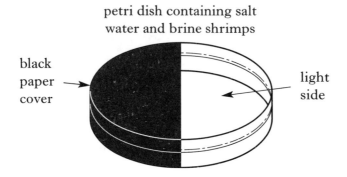

(a) State **one** variable that should be kept constant when setting up the investigation.

_____

(b) Why were the brine shrimps allowed to swim around for two minutes before the investigation was started?

_____

DO NOT WRITE IN THIS MARGIN

Marks

5. (continued)

(*c*) The results are shown in the table below.

	Number of brine shrimps after two minutes	
Student Group	*Dark side*	*Light side*
A	4	2
B	1	5
C	3	3
D	2	4
E	1	5
Total	11	19

(i) From the results describe the response of brine shrimps to light.

____

____ **1**

(ii) Explain why this response helps the brine shrimp survive.

____

____ **1**

(*d*) Suggest **one** way in which the reliability of the results could be improved.

____

____ **1**

[Turn over

6. (*a*) In farmyard fowl, feather type is controlled by a single gene. The allele for normal feathers (N) is **co-dominant** with the allele for extreme frizzle feathers (F). The results of a cross between two homozygous fowl is shown below.

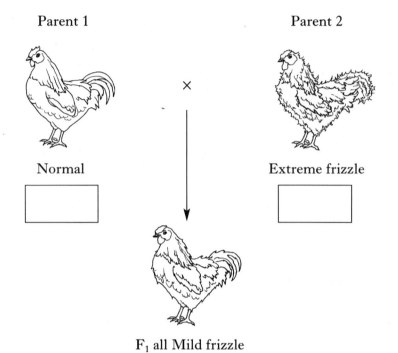

Parent 1 — Normal

Parent 2 — Extreme frizzle

F₁ all Mild frizzle

(i) Complete the blank boxes in the diagram above to show the genotypes of the parents.

(ii) Two mild frizzle fowl from the F₁ were crossed together.

Complete the punnet square below to show the genotype of the gametes from the F₁ male parent and the genotypes of the F₂ produced.

		genotype of gametes from F₁ male parent	
genotype of gametes from F₁ female parent	N		
	F		

(iii) State the expected F₂ phenotype ratio.

Ratio ____ normal : ____ mild frizzle : ____ extreme frizzle

6. (continued)

(b) Complete the table below by writing the correct word from the list to match the description.

List
interbreeding
recessive
heterozygous
homozygous
monohybrid

Description	Word
A genotype with different alleles of a particular gene.	
An allele which is always masked by a dominant allele.	
A type of cross between two true breeding parents that differ in one characteristic.	

3

(c) Skin colour is an example of a human characteristic controlled by the alleles of more than one gene.

What name is given to this type of inheritance?

_____

1

[Turn over

7. The diagram below shows part of a food web found on a rocky shore.

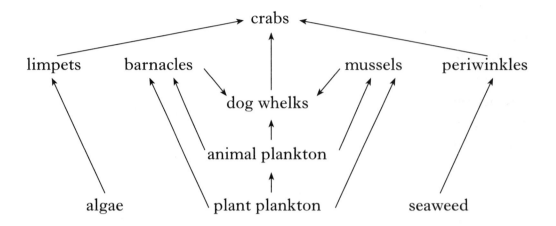

(a) Use the words "increases", "decreases", or "stays the same" to suggest what might happen to the populations of barnacles and periwinkles if all the mussels were removed. Give a reason for each answer.

1 Barnacle population _____

   Reason _____

   _____

   _____ **1**

2 Periwinkle population _____

   Reason _____

   _____

   _____ **1**

(b) Why is the biomass of algae greater than the biomass of limpets?

_____

_____ **1**

7. (continued)

(c) The following diagram shows a pyramid of energy for part of the rocky shore ecosystem.

The energy values are given in kJ/m²/year.

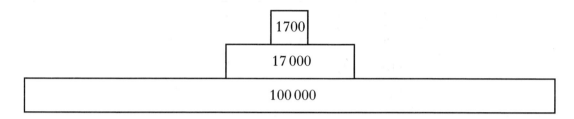

(i) Why does the energy value decrease from one level to the next?

_____

_____   1

(ii) Use information from the food web and the pyramid of energy to complete the table below.

Energy value (kJ/m²/year)	Niche	Named organism
100 000		
	primary consumer	animal plankton
		dog whelks

2

[Turn over

8. (a) The table below gives information about wheat produced by selective breeding over many generations.

Generation number	Average height of stem (cm)	Grain yield (tonnes per hectare)	Average length of grain (mm)
1	142	6·0	10
27	126	6·0	9
45	110	6·7	11
64	106	7·5	11
72	84	8·7	10

From the table, describe **one** improvement in the wheat and explain why it is a desirable characteristic.

Improvement _____

Explanation _____

_____

(b) Give **one** disadvantage of selective breeding.

_____

_____

(c) Genetic engineering can be used to transfer human genes to bacteria.

(i) Name a human hormone which can be produced by genetically engineered bacteria.

_____

8. (c) (continued)

(ii) In the boxes below, describe each of the steps carried out to transfer successfully a human gene to a bacterial cell.

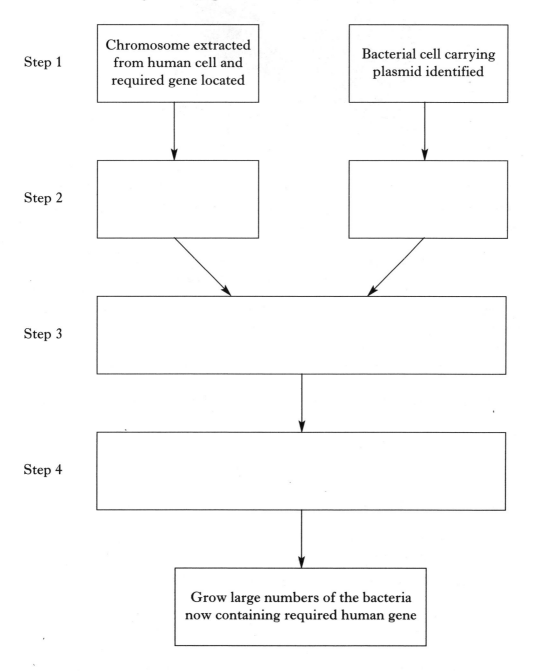

Step 1: Chromosome extracted from human cell and required gene located | Bacterial cell carrying plasmid identified

Step 2:

Step 3:

Step 4:

Grow large numbers of the bacteria now containing required human gene

3

9. The diagram below shows the light and dark varieties of the peppered moth, *Biston betularia*.

Bark of tree

In an investigation moths were captured in a woodland area, marked and released.

Twenty four hours later moths were recaptured and the results are shown in the table below.

Variety	Number of moths marked and released	Number of marked moths recaptured	Percentage recaptured
Light	320	192	60
Dark	280	112	40

(a) (i) Explain why it was necessary to calculate the **percentage** of moths recaptured.

(ii) The results indicate that the investigation was carried out in a non-industrial area.

Explain why the percentage of light coloured moths recaptured is high.

(b) What name is given to the process which results in the difference in numbers of these two varieties in this area?

10. The pulse rate and breathing rate of a student were taken before, during and after a period of exercise. The graph below shows the results obtained.

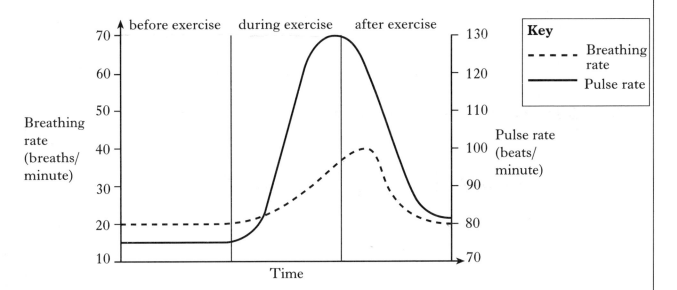

(a) (i) Complete the table to show the changes in pulse rate.

	Before exercise	During exercise	After exercise
Breathing rate (breaths/min)	20	from 20 to 35	from 35 to 40 to 20
Pulse rate (beats/min)			

(ii) Explain why breathing rate increases during the exercise.

_____

_____

(b) Muscle fatigue may occur during exercise. Name the chemical that results in muscle fatigue.

_____

(c) (i) Name the structures in the lungs where gas exchange takes place.

_____

[Turn over

DO NOT WRITE IN THIS MARGIN

Marks

10. (*c*) (continued)

(ii) State **two** ways by which blood carries carbon dioxide around the body.

1 ______________________________

2 ______________________________ **2**

(*d*) Underline **one** option in each set of brackets to make the following sentence correct.

In the lungs haemoglobin { combines with / releases } oxygen at { high / low } oxygen levels. **1**

11. The following diagram shows the human brain.

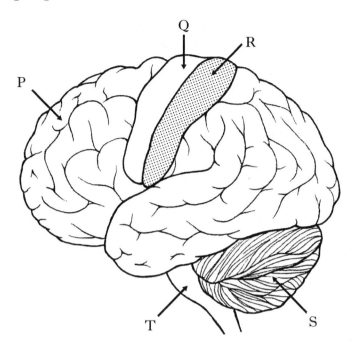

(a) Complete the table to identify areas of the brain and their functions.

Name of area	Letter	Function
Sensory strip		Receives nerve impulses from the sense organs
Cerebellum		
	T	

2

(b) The brain forms one part of the Central Nervous System (CNS).

Name the other part.

_____ **1**

(c) Name the type of neurone which links the receptors in the sense organs to the CNS.

_____ **1**

[Turn over

11. (continued)

(d) Decide if each of the following statements about temperature regulation in the body is **True** or **False**, and tick (✓) the appropriate box.

If the statement is **False**, write the correct word in the **Correction** box to replace the word underlined in the statement.

Statement	True	False	Correction
External temperature is detected by receptors in the <u>skin</u>.			
The area of the brain which regulates body temperature is the <u>medulla</u>.			
Blood vessels in the skin <u>constrict</u> in response to an increase in external temperature.			

3

[Turn over for SECTION C on *Page twenty-eight*]

SECTION C

Both questions in this section should be attempted.

Note that each question contains a choice.

Questions 1 and 2 should be attempted on the blank pages which follow.

Supplementary sheets, if required, may be obtained from the invigilator.

1. Answer **either** A **or** B.

    A. The flow diagram below shows the two stages of photosynthesis.

    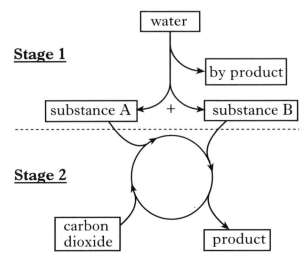

    Name and describe **Stage 1** and **Stage 2**. 5

OR

   B. The diagram below shows a container used for home wine production.

   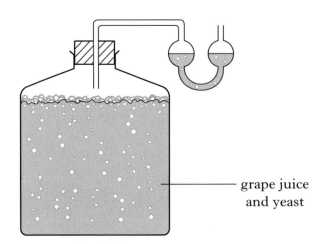

   Describe the anaerobic pathway of respiration which results in wine production in this container. 5

Question 2 is on *Page thirty*.

SPACE FOR ANSWER TO QUESTION 1

2. Answer **either** A **or** B.

Labelled diagrams may be included where appropriate.

A. Describe the structures of arteries, veins and capillaries. Give the function of each of these types of blood vessel. **5**

OR

B. Freshwater bony fish have a water balance problem. State the water balance problem and describe how these fish overcome the problem. **5**

[END OF QUESTION PAPER]

SPACE FOR ANSWER TO QUESTION 2

SPACE FOR ANSWERS

ADDITIONAL GRAPH PAPER FOR QUESTION 4(a)(i)

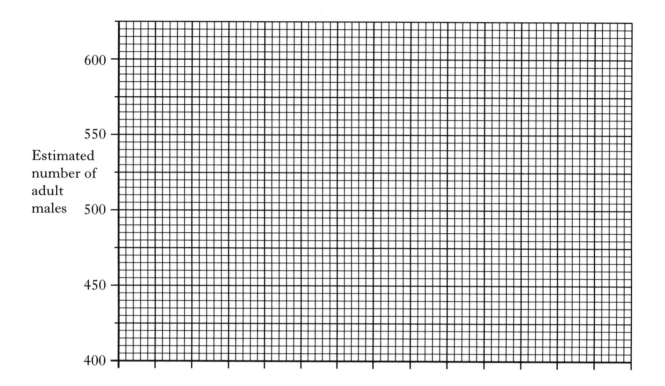

2004 | Intermediate 2

Official SQA Past Papers: Intermediate 2 Biology 2004

FOR OFFICIAL USE

X007/201

Total for Sections B and C

NATIONAL QUALIFICATIONS 2004

WEDNESDAY, 19 MAY 9.00 AM – 11.00 AM

BIOLOGY INTERMEDIATE 2

Fill in these boxes and read what is printed below.

Full name of centre

Town

Forename(s)

Surname

Date of birth
Day Month Year   Scottish candidate number   Number of seat

SECTION A (25 marks)

Instructions for completion of Section A are given on page two.

SECTIONS B AND C (75 marks)

1 (a) All questions should be attempted.

   (b) It should be noted that in **Section C** questions 1 and 2 each contain a choice.

2 The questions may be answered in any order but all answers are to be written in the spaces provided in this answer book, and must be written clearly and legibly in ink.

3 Additional space for answers and rough work will be found at the end of the book. If further space is required, supplementary sheets may be obtained from the invigilator and should be inserted inside the **front** cover of this book.

4 The numbers of questions must be clearly inserted with any answers written in the additional space.

5 Rough work, if any should be necessary, should be written in this book and then scored through when the fair copy has been written.

6 Before leaving the examination room you must give this book to the invigilator. If you do not, you may lose all the marks for this paper.

Official SQA Past Papers: Intermediate 2 Biology 2004

Read carefully

1. Check that the answer sheet provided is for Biology Intermediate 2 (Section A).
2. Fill in the details required on the answer sheet.
3. In this section a question is answered by indicating the choice A, B, C or D by a stroke made in **ink** in the appropriate place in the answer sheet—see the sample question below.
4. For each question there is only **one** correct answer.
5. Rough working, if required, should be done only on this question paper, or on the rough working sheet provided—**not** on the answer sheet.
6. At the end of the examination the answer sheet for Section A **must** be placed inside the front cover of this answer book.

Sample Question

Which part of the brain is involved in the control of heart rate?

A Cerebellum

B Medulla

C Hypothalamus

D Cerebrum

The correct answer is B—Medulla. A **heavy** vertical line should be drawn joining the two dots in the appropriate box in the column headed **B** as shown **in the example on the answer sheet**.

If, after you have recorded your answer, you decide that you have made an error and wish to make a change, you should cancel the original answer and put a vertical stroke in the box you now consider to be correct. Thus, if you want to change an answer **D** to an answer **B**, your answer sheet would look like this:

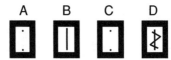

If you want to change back to an answer which has already been scored out, you should **enter a tick (✓)** to the RIGHT of the box of your choice, thus:

SECTION A

All questions in this Section should be attempted.

1. The energy values of different food materials are shown in the table.

Food	Energy value (kJ per gram)
Glucose	4
Protein	4
Fat	9

   How much energy is contained in a food sample consisting of 3 grams of glucose and 2 grams of fat?

   A 17 kJ
   B 21 kJ
   C 30 kJ
   D 35 kJ

2. The function of the villi is to increase the surface area for

   A absorption
   B protection
   C acid production
   D peristalsis.

3. Bile is stored in the

   A liver
   B gall bladder
   C stomach
   D small intestine.

4. A piece of carrot weighs 20 g fresh and 2 g dry. What is the percentage water content of the carrot?

   A 2%
   B 10%
   C 72%
   D 90%

5. The table below shows the rate of blood flow to the body at rest and during strenuous exercise.

   Which line in the table shows the greatest increase in blood flow during strenuous exercise?

	Region of body	Blood flow (cm^3/minute)	
		at rest	strenuous exercise
A	brain	750	750
B	muscle	1200	22 000
C	heart	250	750
D	skin	500	600

Questions 6 and 7 refer to the diagram which shows the structure of the lungs.

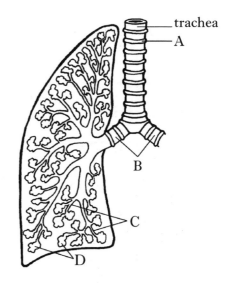

6. Which label identifies the bronchioles?

7. The function of part A is to

   A prevent the lungs from collapsing
   B keep the trachea open at all times
   C prevent food entering the windpipe
   D trap dirt and bacteria.

[Turn over

8. Which line in the table below identifies correctly how macrophages destroy bacteria?

	Phagocytosis	Antibody production
A	yes	yes
B	yes	no
C	no	yes
D	no	no

9. The diagram below represents a unicellular organism.

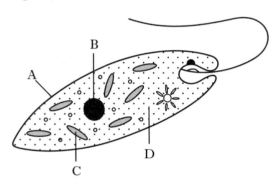

Which part indicates this is a plant cell?

10. The diagram below shows onion cells as observed under a microscope at a magnification of 100 X.

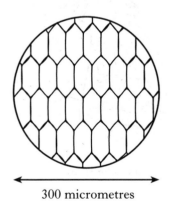

300 micrometres

The diameter of the field of view is 300 micrometres. The average width of each cell in micrometres is

A 0·38

B 0·75

C 37·5

D 75·0.

11. Which line in the table below correctly matches the organism, product and the commercial use of the product?

	Organism	Product	Commercial use of product
A	yeast	methane	biogas
B	bacteria	alcohol	biogas
C	yeast	alcohol	gasohol
D	bacteria	methane	gasohol

12. The graph below shows the effect of increasing antibiotic concentrations on the percentage of bacteria surviving within a population. None of the bacteria had resistance to the antibiotic.

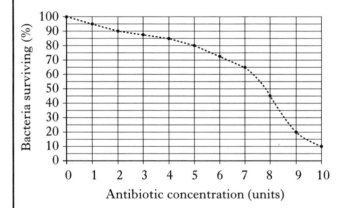

Another experiment was carried out with different bacteria, some of which had resistance to the antibiotic.

Which of the following best describes the effect on the bacteria surviving in this second experiment?

A The percentage of bacteria surviving would increase.

B The percentage of bacteria surviving would decrease.

C There would be no change in the percentage of bacteria surviving.

D All of the bacteria would survive.

13. Two grams of fresh liver was added to hydrogen peroxide.

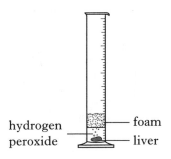

The time taken to collect 10 cm³ of oxygen foam was 2 minutes.

The rate of oxygen production was

A 2·5 cm³/g/min
B 5·0 cm³/g/min
C 10·0 cm³/g/min
D 20·0 cm³/g/min.

14. The diagram below illustrates an investigation of respiration in yeast.

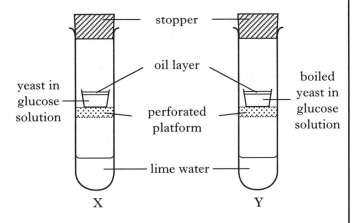

Lime water is an indicator which changes from clear to cloudy in the presence of carbon dioxide.

The investigation was allowed to run for 24 hours.

Which line in the table below identifies correctly the appearance of the lime water in tubes X and Y after 24 hours?

	X	Y
A	clear	clear
B	cloudy	cloudy
C	clear	cloudy
D	cloudy	clear

15. Which of the following are **all** limiting factors in photosynthesis?

A Carbon dioxide concentration, temperature and light intensity

B Carbon dioxide concentration, oxygen concentration and light intensity

C Oxygen concentration, temperature and light intensity

D Oxygen concentration, carbon dioxide concentration and temperature

16. Which line in the table below identifies the **best** conditions for the production of early crops?

	Added factor	Light intensity
A	oxygen	high
B	oxygen	medium
C	carbon dioxide	medium
D	carbon dioxide	high

17. The following stages occur during photosynthesis.

W glucose is formed
X water is broken down to produce hydrogen
Y glucose is converted to starch
Z hydrogen is combined with carbon dioxide

The correct order for these stages is

A W Z X Y
B Z Y X W
C X Z W Y
D Y X Z W.

18. Which of the following is a correct description of a decomposer?

A A micro-organism which lives inside animals and causes disease.

B An organism which releases chemicals from organic waste.

C A fungus which grows on living tissue.

D A green plant which roots in rotting vegetation.

[Turn over

19. The following diagram shows a pyramid of energy. Which level is the result of the energy from the sun being converted into chemical energy?

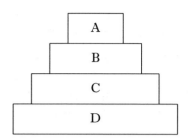

20. The following choice chamber was used to investigate the effect of humidity on the behaviour of woodlice.

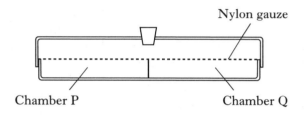

Which line in the table below describes the **best** experimental set up?

	Number of woodlice	Contents of chamber P	Contents of chamber Q	Modification to choice chamber
A	10	Drying agent	Wet cotton wool	Half covered in black paper
B	10	Wet cotton wool	Drying agent	Totally covered in black paper
C	20	Drying agent	Wet cotton wool	Half covered in black paper
D	20	Wet cotton wool	Drying agent	Totally covered in black paper

21. The diagram below shows the main parts of a flower.

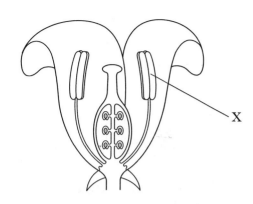

Which line in the table identifies X and the type of gamete it produces?

	Name of X	Type of gamete produced
A	ovary	male
B	ovary	female
C	anther	female
D	anther	male

22. The information below refers to some woodland birds.

Bird species	Common food eaten	Nest location
Lesser spotted woodpecker	insects	dead trees
Green woodpecker	ants, other insects	live trees
Greater spotted woodpecker	insects, nuts, seeds	live trees
Treecreeper	insects, spiders, seeds	dead trees

Between which two bird species will competition for food and nest location be greatest?

A Lesser spotted woodpecker and treecreeper

B Greater spotted woodpecker and lesser spotted woodpecker

C Lesser spotted woodpecker and green woodpecker

D Greater spotted woodpecker and treecreeper

23. In humans, all sperm contain

A an X chromosome

B a Y chromosome

C an X and Y chromosome

D either an X or a Y chromosome.

24. The graph below shows the average number of peppered moths, in a woodland, in June of each year over a 10 year period.

Key ----- light form
—■— dark form

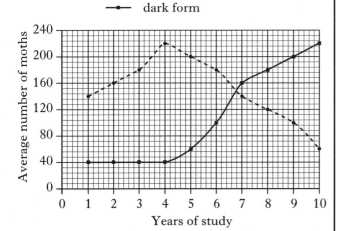

Studies have shown that an increase in the number of dark moths is related to an increase in the level of pollution in the atmosphere.

Which of the following **best** describes what would happen to the number of moths if measures were introduced to reduce air pollution from year 7?

A Increase in dark moths and decrease in light moths

B Decrease in dark moths and increase in light moths

C Increase in dark moths and increase in light moths

D Decrease in dark moths and decrease in light moths

25. Genetic engineering can be used to alter bacterial cells in order to produce human insulin.

The following stages occur during genetic engineering.

1 Insulin gene extracted from a human cell

2 Bacteria divide and produce large quantities of human insulin

3 Plasmid is removed from bacterial cell and "cut" open

4 Insulin gene is inserted into bacterial plasmid

The correct sequence of these stages is

A 1 3 4 2

B 1 3 2 4

C 3 4 2 1

D 3 1 2 4.

Candidates are reminded that the answer sheet for Section A MUST be placed INSIDE the front cover of this answer book.

[Turn over for Section B on *Page eight*

SECTION B

All questions in this section should be attempted.

1. An experiment was set up to investigate the effect of pH on the action of the enzyme salivary amylase.

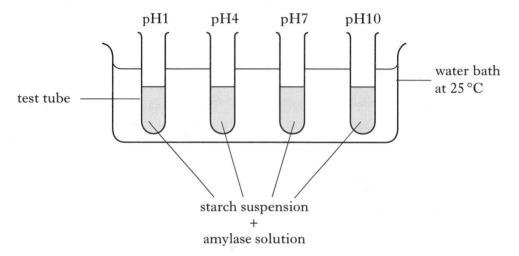

After 30 minutes a sample from each test tube was tested for the presence of simple sugars.

(a) (i) Other than temperature, state **two** variables that must be kept constant in the experiment.

1 _____

2 _____   2

(ii) Name the reagent used to test for simple sugars.

_____   1

(b) The results obtained are shown in the table below.

pH	Simple sugars test
1	negative
4	negative
7	positive
10	negative

DO NOT WRITE IN THIS MARGIN

Marks

1. (*b*) (continued)

(i) What conclusion can be drawn from these results?

____

____ **1**

(ii) Predict the results if the enzyme had been boiled before use. Give an explanation for your answer.

Prediction ____________________________________

Explanation ____________________________________

____ **2**

(*c*) Explain why food containing starch must be digested before it can be used in the human body.

____

____

____ **2**

[Turn over

Marks

2. (a) Complete the following sentences by <u>underlining</u> **one** option in each pair of brackets to describe correctly the body's response to exposure to **low** temperature.

The temperature change is detected by receptors in the skin which send nerve impulses to the { hypothalamus / pituitary }. Nerve impulses are then sent to arterioles in the skin causing them to { constrict / dilate }. Sweat production { decreases / increases } to help return the body temperature to normal.   **2**

(b) The table below lists the stages in a reflex arc. Each stage is represented by a letter.

Stage	Letter
An impulse passes through a motor neurone	A
An impulse passes through a sensory neurone	B
The effector brings about a response	C
A receptor detects a stimulus	D
An impulse passes through a relay neurone	E

(i) Complete the following flow chart to show the correct order of these stages.
The first stage has been given.

1

(ii) What is the function of reflex actions?

_____   **1**

[X007/201] Page ten

3. The diagram shows part of the digestive system.

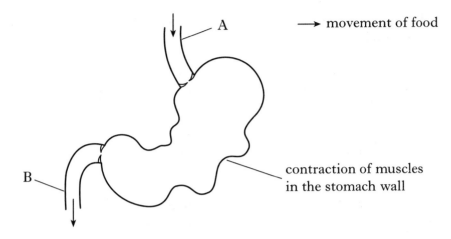

(a) Name structures A and B.

A _____

B _____ 2

(b) Name **one** type of muscle found in the stomach wall.

_____ 1

(c) How do the contractions of the muscles in the stomach wall help the digestion of food?

_____ 1

[Turn over

4. The oxygen concentration of the air decreases as the height above sea level increases.

The table below shows the red blood cell count of a mountaineer taken at different heights above sea level.

Height above sea level (metres)	Red blood cell count (millions/mm³ of blood)
200	5·0
1000	5·6
2200	6·5
3600	7·6
4800	8·5

(a) On the grid, plot a line graph to show red blood cell count against height above sea level.

(Additional graph paper, if required, will be found on page 30.)

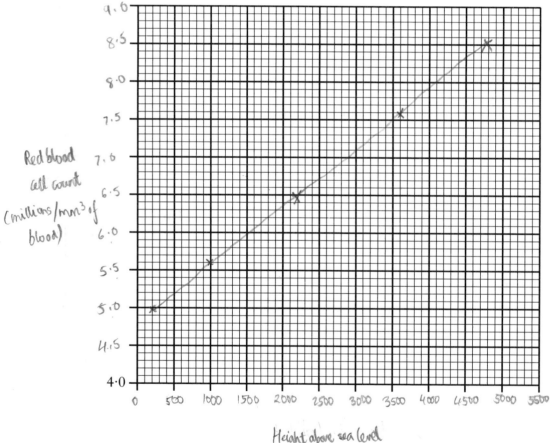

DO NOT WRITE IN THIS MARGIN

Marks

4. (continued)

(*b*) (i) From the table, describe the relationship between the height above sea level and the red blood cell count.

____

____ **1**

(ii) Explain the importance of this change in the red blood cell count.

____

____ **1**

[Turn over

Marks

5. (a) The diagram below shows the unicellular organism *Paramecium* which lives in freshwater.

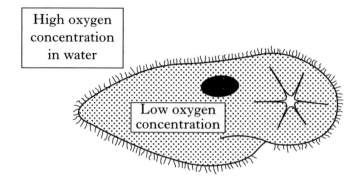

(i) Name the process by which oxygen moves from the water into the organism.

_____   1

(ii) Name a substance that moves from the organism into the water.

_____   1

(iii) Name the cell structure which controls the entry and exit of materials.

_____   1

5. (continued)

(b) The diagram below shows the internal structure of a leaf.

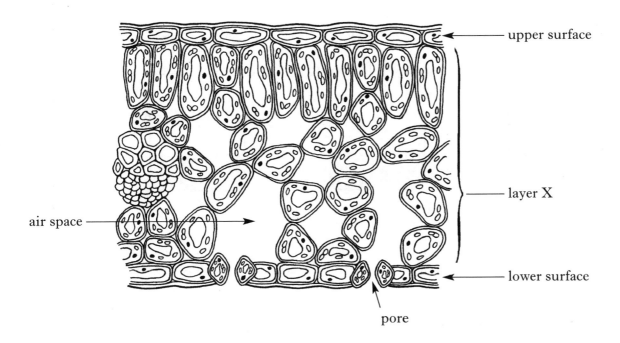

During the hours of daylight, the concentration of carbon dioxide in the air spaces is higher than in the cells of layer X.

Explain why this difference in concentration is important to the leaf cells.

_____

_____

_____   2

[Turn over

6. Four groups of students carried out an investigation into the effect of the micro-organisms present in live yoghurt on milk. The milk was first sterilised to remove all micro-organisms present. Each group set up 3 beakers as shown below.

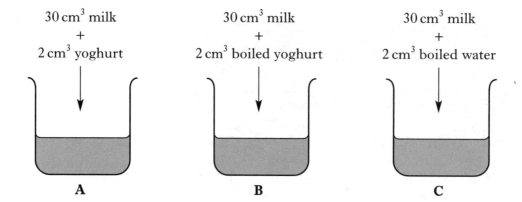

A lid was placed over each beaker. The beakers were placed in a water bath at 35 °C. The pH of the contents of each beaker was measured at the start and 6 hours later.

(a) (i) Name the type of micro-organism found in live yoghurt.

_____ 1

(ii) Name the substance, produced by these micro-organisms, which causes a change in the pH of the milk.

_____ 1

(iii) Suggest why the lid was placed over each beaker.

_____

_____ 1

(iv) Why was beaker C included as a control?

_____

_____ 1

6. (continued)

(b) The results from the 4 groups are given in the table below.

Beaker	Change in pH				
	Group 1	Group 2	Group 3	Group 4	Average
A	−1·3	−1·8	−1·0	−1·5	
B	0·0	−1·2	0·0	0·0	−0·3
C	0·0	0·0	0·0	0·0	0·0

(i) Complete the table to show the average change in pH for beaker A.
Space for calculation

(ii) Why were the results from the 4 groups collected and an average calculated?

(iii) Account for the unexpected result in beaker B of group 2.

[Turn over

7. (a) An investigation into the effects of solutions of different salt concentrations on red blood cells was carried out. Three microscope slides were set up as shown below.

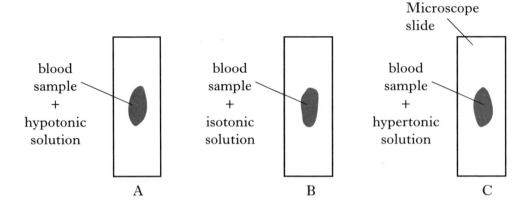

Each blood sample was observed under a microscope after 10 minutes.

(i) Describe what would have happened to the red blood cells on slides A and C.

Slide A _____

Slide C _____   1

(ii) Name the process responsible for these changes.

_____   1

(iii) What is meant by an **isotonic solution**?

_____

_____   1

(b) (i) State **one** osmoregulatory problem experienced by marine bony fish.

_____   1

(ii) Describe **one** method used by these fish to overcome the problem.

_____   1

8. The diagram below shows the main stages of aerobic respiration.

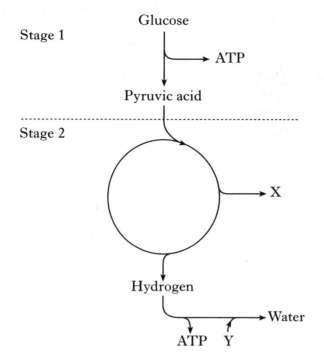

(a) Name Stage 1.

_____

(b) Name substances X and Y.

X _____

Y _____

(c) Which substance shown in the diagram is the **source** of the energy used to synthesise ATP?

_____

(d) Complete the following word equation which represents the synthesis of ATP.

_____ + _____ + energy ⟶ ATP

(e) How many molecules of ATP are produced per glucose molecule during each of the following stages?

Stage 1 _____

Stage 2 _____

(f) During aerobic respiration some energy is lost from the cell.
In what form is this energy?

_____

9. (a) A natural pine forest provides excellent habitats for many different organisms.

One of these organisms is a large bird called the capercaillie which nests in the deep vegetation on the forest floor. In summertime it eats berries, leaves and stems of the blaeberry and other forest plants. In winter it eats Scots pine needles and cones.

The capercaillie's natural predators are the fox and the wild cat. Crows eat their eggs.

Use information from the passage to answer the following questions.

(i) Complete the boxes below to show a food chain.

(ii) Complete the table of terms and named examples from the passage.

Term	Named example
ecosystem	
	all the crows
herbivore	

(iii) Natural pine forests show high biodiversity. What is meant by the term biodiversity?

_____

_____

DO NOT WRITE IN THIS MARGIN

Marks

9. (continued)

(*b*) The number of capercaillie in Scotland fell from 20 000 in 1970 to 3000 in 1991.

During the same period there was a large increase in the numbers of animals such as deer and sheep which graze on the forest floor.

Explain how this might have caused the decrease in the numbers of capercaillie.

____

____ **1**

(*c*) Give **one** example of a human activity which could affect biodiversity.

____ **1**

[Turn over

Marks

10. (*a*) In the fruitfly *Drosophila melanogaster*, the dominant form (G) of one gene determines grey body colour; black body colour results from the recessive form (g) of the gene.

The genotypes of the parent flies used in a cross are shown below.

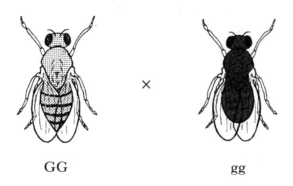

GG   gg

(i) State the genotype(s) of the F_1 offspring of this cross.

_____   **1**

(ii) Decide if each of the following statements about this cross is **True** or **False**, and tick (✓) the appropriate box.

If the statement is **False**, write the correct word in the **Correction** box to replace the word underlined in the statement.

Statement	True	False	Correction
The different forms of the gene are <u>hybrids</u>.			
The <u>parents</u> in this cross are true breeding.			
The F_1 flies are <u>homozygous</u>.			

3

10. **(a)** **(continued)**

(iii) Two flies from the F_1 were allowed to breed together. This produced 56 grey flies and 14 black flies in the F_2.
Express this result as a simple whole number ratio.
Space for calculation

_____ grey flies : _____ black flies

(iv) The expected ratio of grey flies to black flies in the F_2 is 3:1. Suggest why the observed ratio was different from the expected ratio.

_____

_____

(b) In a study of variation, a group of students collected information on the heights and blood groups of a class.
For each variation state whether it is continuous or discontinuous.

Height _____

Blood groups _____

(c) Polygenic inheritance occurs as a result of the interaction of several genes.
Give an example of polygenic inheritance in humans.

_____

[Turn over

11. (a) Complete the table to give the site of production and number of chromosomes of each type of gamete.

Human gamete	Site of production	Number of chromosomes
egg		
sperm		

2

(b) The diagram below shows the chromosome complement of a cell about to divide to form gametes.

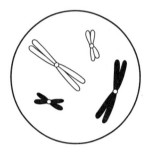

(i) How many sets of chromosomes does this cell contain?

_____

1

(ii) Name the type of cell division which produces gametes.

_____

1

(iii) The following diagram shows one way in which these chromosomes may line up during cell division.

Complete the diagram below to show one other way in which the chromosomes may line up.

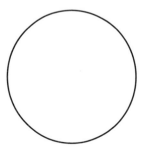

1

11. (continued)

(c) Chromosome pairs line up in a variety of ways.
Explain why this random assortment is important.

_____

_____ **1**

[Turn over for SECTION C on *Page twenty-six*

SECTION C

Both questions in this section should be attempted.

Note that each question contains a choice.

Questions 1 and 2 should be attempted on the blank pages which follow.

Supplementary sheets, if required, may be obtained from the invigilator.

1. Answer **either** A **or** B.

   A. The diagram below shows a section through the human heart.

   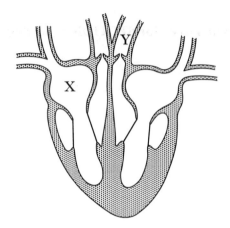

   Describe the pathway of blood through the heart and associated structures starting at X and finishing at Y. There is no need to mention the valves. **5**

   OR

   B. Urine production occurs in the kidney. The diagram below shows the structure of a nephron and its blood supply.

   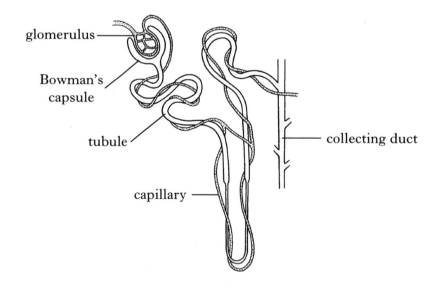

   Describe how the nephron produces urine. There is no need to mention the role of ADH. **5**

   Question 2 is on *Page twenty-eight*.

SPACE FOR ANSWER TO QUESTION 1

2. Answer **either** A **or** B.

   Labelled diagrams may be included where appropriate.

   A. Plants living in the desert are adapted for survival. Describe **three** adaptations and explain how each adaptation increases the chances of survival of the plant. **5**

   OR

   B. Describe the structure of chromosomes. Explain how chromosomes determine the characteristics of an organism. **5**

   [END OF QUESTION PAPER]

SPACE FOR ANSWER TO QUESTION 2

SPACE FOR ANSWERS

ADDITIONAL GRAPH PAPER FOR QUESTION 4(a)

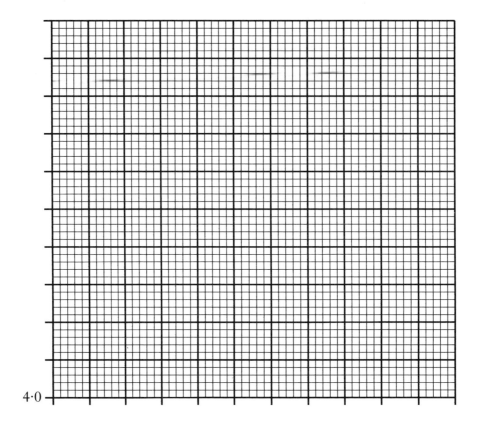

2005 | Intermediate 2

FOR OFFICIAL USE

Total for Sections B and C

X007/201

NATIONAL QUALIFICATIONS 2005

WEDNESDAY, 18 MAY 9.00 AM – 11.00 AM

BIOLOGY INTERMEDIATE 2

Fill in these boxes and read what is printed below.

Full name of centre

Town

Forename(s)

Surname

Date of birth
Day Month Year

Scottish candidate number

Number of seat

SECTION A (25 marks)
Instructions for completion of Section A are given on page two.

SECTIONS B AND C (75 marks)

1 (a) All questions should be attempted.

   (b) It should be noted that in **Section C** questions 1 and 2 each contain a choice.

2 The questions may be answered in any order but all answers are to be written in the spaces provided in this answer book, and must be written clearly and legibly in ink.

3 Additional space for answers will be found at the end of the book. If further space is required, supplementary sheets may be obtained from the invigilator and should be inserted inside the **front** cover of this book.

4 The numbers of questions must be clearly inserted with any answers written in the additional space.

5 Rough work, if any should be necessary, should be written in this book and then scored through when the fair copy has been written. If further space is required, a supplementary sheet for rough work may be obtained from the invigilator.

6 Before leaving the examination room you must give this book to the invigilator. If you do not, you may lose all the marks for this paper.

Read carefully

1. Check that the answer sheet provided is for **Biology Intermediate 2 (Section A)**.
2. Check that the answer sheet you have been given has **your name**, **date of birth**, **SCN** (Scottish Candidate Number) and **Centre Name** printed on it.

   Do not change any of these details.
3. If any of this information is wrong, tell the Invigilator immediately.
4. If this information is correct, **print** your name and seat number in the boxes provided.
5. Use **black** or **blue ink** for your answers. **Do not use red ink**.
6. The answer to each question is **either** A, B, C or D. Decide what your answer is, then put a horizontal line in the space provided (see sample question below).
7. There is **only one correct** answer to each question.
8. Any rough working should be done on the question paper or the rough working sheet, **not** on your answer sheet.
9. At the end of the exam, put the **answer sheet for Section A inside the front cover of this answer book**.

Sample Question

What must be present in leaf cells for photosynthesis to take place?

A   Oxygen and water

B   Carbon dioxide and water

C   Carbon dioxide and oxygen

D   Oxygen and hydrogen

The correct answer is **B**—Carbon dioxide and water. The answer **B** has been clearly marked with a horizontal line (see below).

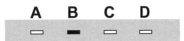

Changing an answer

If you decide to change your answer, cancel your first answer by putting a cross through it (see below) and fill in the answer you want. The answer below has been changed to **B**.

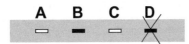

If you then decide to change back to an answer you have already scored out, put a tick (✓) to the **right** of the answer you want, as shown below:

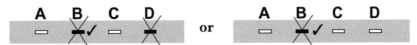

SECTION A

All questions in this Section should be attempted.

1. The diagram below represents a plant cell.

   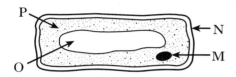

   Which of the labelled parts of the cell are also found in an animal cell?

   A M and N
   B N and O
   C M and P
   D M, N, O and P

2. Which line in the table below describes correctly the functions of the cell wall and chloroplasts in plant cells?

	Function of cell wall	Function of chloroplast
A	prevents cell bursting	respiration
B	controls entry of substances	respiration
C	prevents cell bursting	photosynthesis
D	controls entry of substances	photosynthesis

3. When animal cells are placed in a hypotonic solution they

   A remain unchanged
   B burst
   C plasmolyse
   D become turgid.

4. A piece of potato was cut from a potato tuber and weighed. It was placed in pure water for an hour then removed, dried and weighed again. Finally, it was placed in a concentrated sugar solution for an hour, removed, dried and weighed again.

   Which line in the table records the results most likely obtained by this treatment?

	First weighing	Second weighing	Third weighing
A	5 g	6 g	4 g
B	5 g	4 g	6 g
C	6 g	5 g	4 g
D	5 g	4 g	3 g

5. The anaerobic respiration of one molecule of glucose results in the net gain of

   A 2 molecules of ATP
   B 2 molecules of ADP
   C 38 molecules of ATP
   D 38 molecules of ADP.

[Turn over

6. The graphs below show the effects of temperature and pH on the activity of an enzyme.

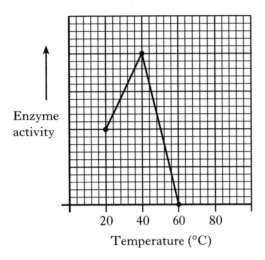

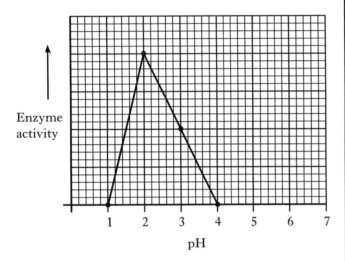

Which line in the table identifies correctly the conditions at which the enzyme is most active?

	Temperature	pH
A	40	2
B	40	4
C	50	2
D	60	4

7. The diagram below shows the respiratory pathway in an animal cell.

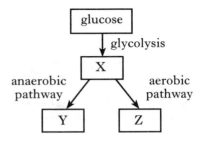

Which line in the table below identifies correctly X, Y and Z?

	X	Y	Z
A	lactic acid	pyruvic acid	carbon dioxide and water
B	carbon dioxide and water	pyruvic acid	lactic acid
C	pyruvic acid	carbon dioxide and water	lactic acid
D	pyruvic acid	lactic acid	carbon dioxide and water

8. Photolysis is the

A combining of water with carbon dioxide

B use of water by chlorophyll to split light

C release of energy from water using light energy

D splitting of water using light energy.

9. ATP synthesised during photolysis provides the carbon fixation stage of photosynthesis with

A glucose

B carbon dioxide

C energy

D hydrogen.

10. Which of the following describes a community?

   A The total number of one species present
   B All the living organisms and the non-living parts
   C All the living organisms
   D All the plants

11. The bar chart shows the results of a survey into the heights of bell heather plants on an area of moorland.

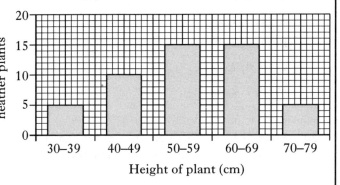

   The percentage of plants with a height greater than 59 cm is

   A 15%
   B 20%
   C 30%
   D 40%.

12. A survey was carried out on numbers of mussels attached to rocks on a sea shore.

   Squares measuring 10 cm × 10 cm were used in the survey.

   The positions of the squares and the number of mussels in each square are shown below.

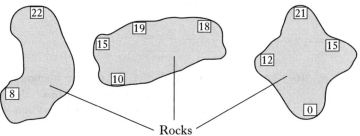

   How could the results have been made more valid?

   A Sample only one rock
   B Use bigger squares
   C Note all species present
   D Count each at the same time of day

13. Plants convert 1% of the light energy they receive into new plant material.

   In the food chain below, plant plankton receive 100 000 units of light energy.

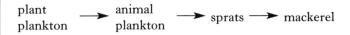

   How much of this energy is converted into new plant material?

   A 10 000 units
   B 1000 units
   C 100 units
   D 10 units

14. The DNA of a chromosome carries information which determines the structure and function of

   A fats
   B bases
   C carbohydrates
   D proteins.

[Turn over

15. A true breeding red bull is mated with a true breeding white cow. The offspring are all intermediate in colour (roan).

This type of inheritance is

A polygenic

B recessive

C co-dominant

D dominant.

16. In 1997, the USA planted 8·2 million hectares of land with genetically engineered crops. By 1998, this had increased to 20·5 million hectares.

What was the percentage increase in the area sown between 1997 and 1998?

A 12·3%

B 66%

C 150%

D 166·7%

17. In tomato plants, the allele for red fruit is dominant to the allele for yellow fruit.

If a heterozygous tomato plant is crossed with a plant which produces yellow fruit, the expected phenotype ratio of the offspring would be

A 3 red : 1 yellow

B 1 red : 3 yellow

C 1 red : 2 yellow

D 1 red : 1 yellow.

18. *Achoo syndrome* is a dominant characteristic in humans which causes the sufferer to sneeze in response to bright light.

A woman who is homozygous for the syndrome and a man who is unaffected have children.

What proportion of their children would be expected to have *Achoo syndrome*?

A 0%

B 25%

C 50%

D 100%

19. Genetic engineering can be used to alter bacterial cells in order to produce human insulin.

The stages in the process are:

1 insulin gene extracted from a human cell

2 bacteria divide and produce large quantities of human insulin

3 plasmid is removed from bacterial cell and "cut" open

4 insulin gene is inserted into bacterial plasmid.

The correct sequence of these stages is

A 1, 3, 4, 2

B 1, 3, 2, 4

C 3, 4, 2, 1

D 3, 1, 2, 4.

20. Food tests were carried out on different food samples. The results are shown below.

Food sample	Food Tests			
	Starch	Glucose	Protein	Fat
A	positive	negative	positive	positive
B	negative	positive	positive	positive
C	positive	negative	negative	positive
D	positive	positive	negative	negative

Which food sample left a translucent spot on filter paper and also turned brick red when heated with Benedicts solution?

21. Which food group contains the most energy per gram?

A Carbohydrate

B Protein

C Fat

D Vitamins

22. Stomach muscles relax and contract in order to

A release enzymes

B aid absorption of digested products

C release mucus and acid

D mix food with digestive juices.

23. The graph below shows the relationship between oxygen concentration and the concentration of oxyhaemoglobin.

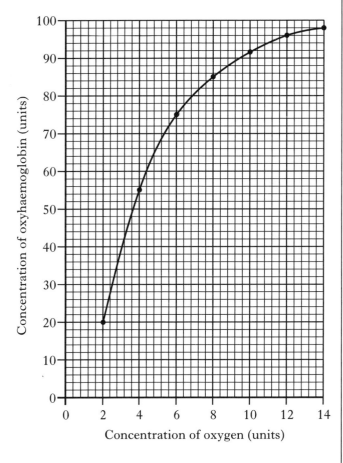

What is the percentage increase in the concentration of oxyhaemoglobin when the concentration of oxygen increases from 6 units to 12 units?

A 6

B 21

C 28

D 96

24. The diagram below shows a human brain.

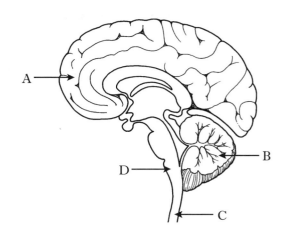

Which letter indicates the site of memory and conscious responses?

25. The diagram below shows the times taken in milliseconds (ms) for nerve impulses to travel along parts of the nervous system.

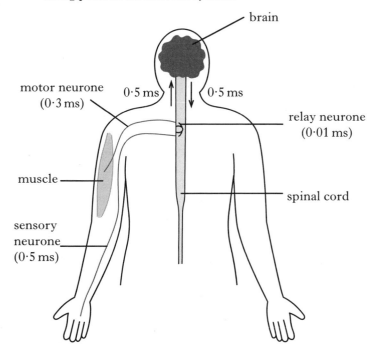

The time taken for a reflex response involving the nerves above is

A 0·81 ms

B 1·01 ms

C 1·80 ms

D 1·81 ms.

Candidates are reminded that the answer sheet for Section A MUST be placed INSIDE the front cover of this answer book.

[Turn over for Section B on *Page eight*

SECTION B

All questions in this section should be attempted.

1. The diagram below shows the human alimentary canal.

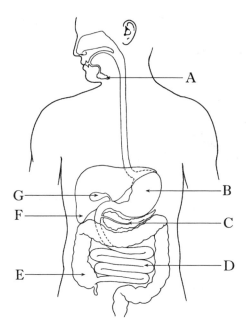

   (a) Name the following labelled parts.

Letter	Name
A	
G	
E	

   (b) Use a letter from the diagram to identify where each of the following secretions are **produced**.

Secretion	Letter
bile	
hydrochloric acid	
lipase	

   (c) Excess glucose in the diet is converted into an insoluble compound which is stored in the liver. Name this compound.

   _____

2. (a) The diagram below shows part of the human urinary system.

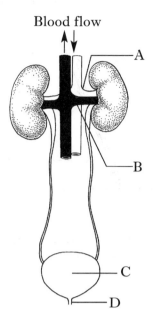

(i) Complete the table below to name the labelled parts and give their functions.

Letter	Name	Function
A	Renal artery	
C		
D		Carries urine out of the body.

(ii) Give **one** difference between the composition of blood in vessels A and B.

_____

(b) Glucose is present in the blood entering the kidney. Explain why glucose does not normally appear in the urine.

_____

_____

(c) (i) Name the hormone which is produced in response to a reduction in water concentration of the blood.

_____

(ii) State the effect this hormone has on the kidney tubules.

_____

3. The diagram below shows the apparatus used to investigate the energy content of different foods. One gram of each food was burned under a beaker containing $100\,cm^3$ of water.

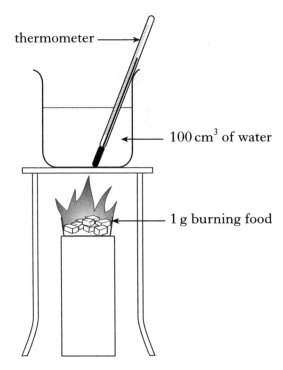

The temperature rise for each food was recorded.
The energy content of the foods was calculated using the following equation.

energy content = temperature rise × 420 (joules/g)

The table below shows the results for the investigation.

Food	Energy Content (joules/g)
butter	10 500
chicken	4200
bread	3400
margarine	10 500

(a) Calculate the **simple whole number ratio** of the energy content of chicken to that of butter.
Space for calculation

_____ : _____
chicken   butter

3. (continued)

(b) Construct a bar graph of the results given in the table.
(Additional graph paper, if required, will be found on page 26.)

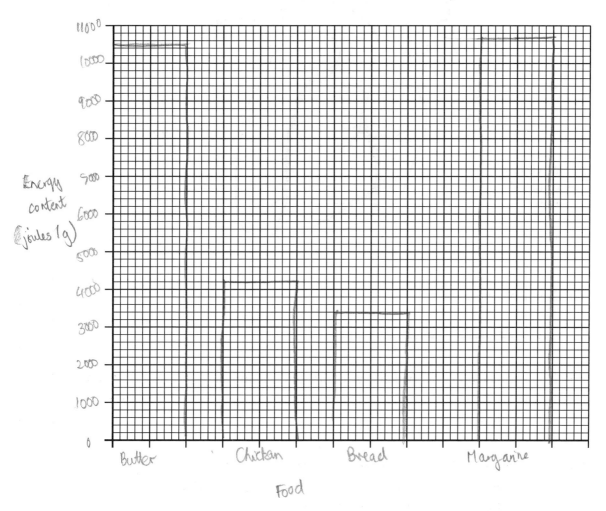

3

(c) One gram of fish was also burned. The temperature rise was 7·5 °C. Calculate the energy content for fish using the equation above.

Space for calculation

Energy content = _____ joules/g **1**

(d) Slimmers may be advised to use margarine instead of butter. Use the data in the table to suggest why this would not aid weight loss.

_____ **1**

4. (a) The diagram below shows a surface view of the human heart.

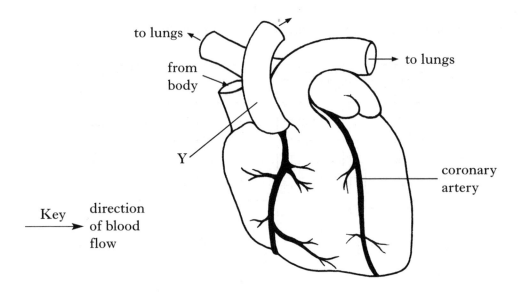

(i) Name blood vessel Y.

_____ 1

(ii) Name the blood vessel which carries blood to the lungs.

_____ 1

(iii) If the coronary artery is blocked, the heart cannot function efficiently.

Name **two** essential substances carried by the blood which would be prevented from reaching the heart muscle.

1 _____

2 _____ 2

4. (continued)

(b) The diagram below shows a type of blood cell which produces antibodies against disease-causing organisms.

(i) Name this type of blood cell.

_____ **1**

(ii) Explain why each antibody is effective against only one type of disease-causing organism.

_____

_____ **1**

(iii) These blood cells produce antibodies when injections are given to protect against disease such as tetanus. Two injections may be given several weeks apart.

The following graphs show the antibody production in response to the two injections.

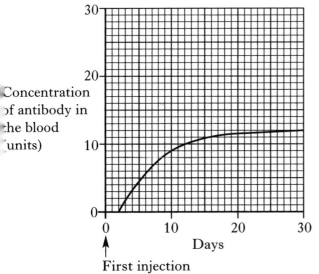
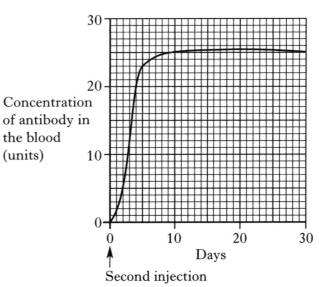

Give **two** differences in the antibody production in response to the two injections.

1 _____

2 _____ **2**

5. (a) The diagram below shows an air sac and a capillary in the lungs.

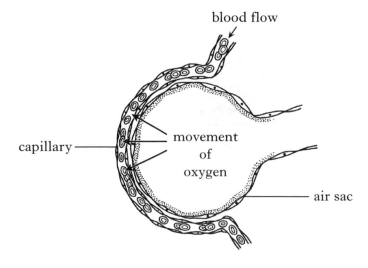

(i) Name the process by which oxygen moves from the air sac into the capillary.

_____ 1

(ii) Why is oxygen required by an organism?

_____ 1

(iii) Complete the following sentence using the words "high" or "low".

Oxygen is moving from a _____ concentration in the air sac to a

_____ concentration in the capillary. 1

(iv) Name a substance which moves from the capillary into the air sac.

_____ 1

(b) Give **two** features of the air sacs which make them efficient gas exchange surfaces.

Feature 1 _____ 1

Feature 2 _____ 1

DO NOT WRITE IN THIS MARGIN

Marks

6. (*a*) Enzymes are involved in synthesis or degradation chemical reactions. The diagram below represents an example of one of these types of reactions.

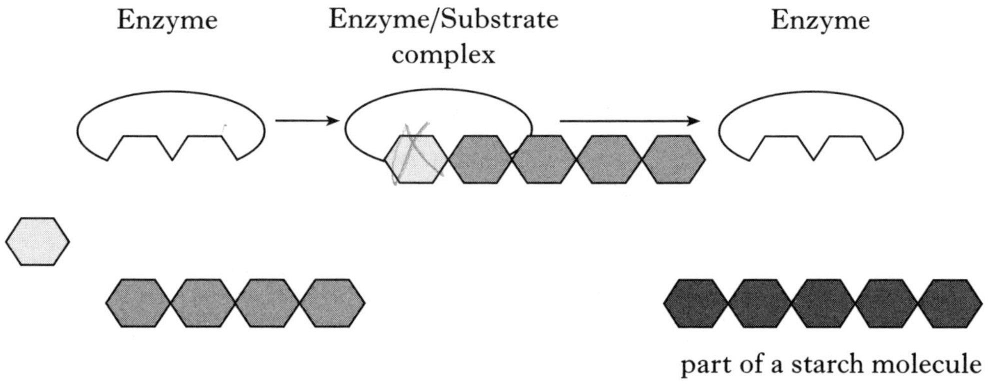

Enzyme
Enzyme/Substrate complex
Enzyme
part of a starch molecule

(i) Name the type of chemical reaction and the enzyme shown in the diagram.

Type of chemical reaction ______________________

The enzyme ______________________ **2**

(ii) Place an X on the diagram to show the position of an active site. **1**

(*b*) What type of molecule are all enzymes made of?

______________________ **1**

(*c*) What happens to the active site when an enzyme is denatured?

______________________ **1**

(*d*) State the effect of an enzyme on the energy input needed for a chemical reaction.

______________________ **1**

[Turn over

7. (a) The diagram below shows part of a food web in a freshwater ecosystem.

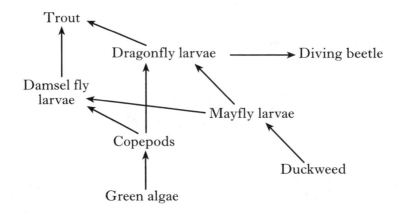

(i) Use **four** organisms from the food web to construct a food chain below.

_____ → _____ → _____ → _____

(ii) Identify **all** the primary consumers in this food web.

_____

(iii) Draw and label a pyramid of numbers from the food web.

(b) What is meant by the term omnivore?

_____

(c) What term is used to describe the variety of species within an ecosystem?

_____

8. (a) The line graph below shows the decomposition of leaves in soil at different temperatures.

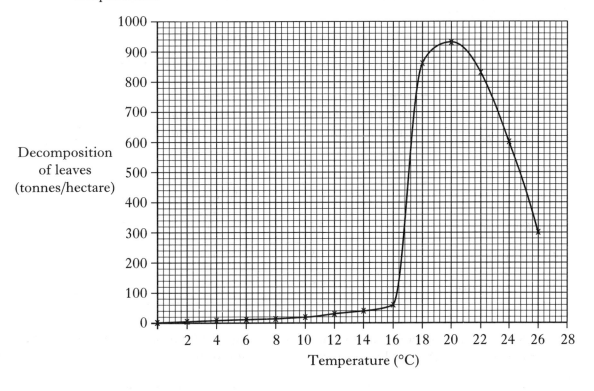

(i) How many times greater is the decomposition of leaves at 24 °C than at 14 °C?

_____ 1

(ii) Describe the relationship between temperature and the decomposition of leaves.

_____
_____ 2

(b) Explain why temperature has an effect on the decomposition of leaves.

_____
_____ 1

(c) (i) Name **one** type of decomposer.

_____ 1

(ii) Describe the role of decomposers in the soil.

_____
_____ 1

Marks

9. (*a*) The table below shows the results of a study into the phenotypes of two pairs of human adult identical twins. Identical twins were used in this study as they have the same genotype.

One pair of identical twins had been raised together since birth.

The second pair had been separated since birth and raised by different families.

Phenotype	Appearance of twins raised together		Appearance of twins raised apart	
	P	Q	R	S
Eye colour	blue	blue	brown	brown
Height (cm)	175	174	180	176
Blood group	A	A	O	O
Hand span (cm)	23	23·5	25	23

From the results, complete the following table by using tick(s) to show whether each phenotype was affected by genes, the environment or both.

Phenotype	Affected by genes	Affected by environment
Eye colour		
Height		
Blood group		
Hand span		

2

(*b*) In another study into plant phenotypes, leaf lengths were found to vary across a wide range.

What term is used to describe this type of variation?

_____

1

9. (continued)

(c) The diagram below shows all the chromosomes found in a human skin cell.

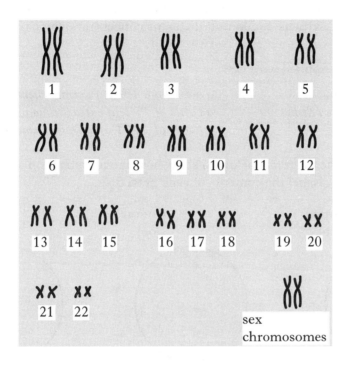

Identify the sex of the person and give a reason for your answer.

Sex _____

Reason _____

_____ 1

(d) Underline **one** option in each set of brackets to make the following sentences correct.

During meiosis, matching chromosomes pair and separate producing $\begin{Bmatrix} \text{gametes} \\ \text{body cells} \end{Bmatrix}$

with $\begin{Bmatrix} \text{one set} \\ \text{two sets} \end{Bmatrix}$ of chromosomes. A zygote is produced from these cells

by $\begin{Bmatrix} \text{random assortment} \\ \text{fertilisation} \end{Bmatrix}$. 2

[Turn over

10. The leaves of black walnut trees produce a chemical which is released into the soil when the leaves fall. This chemical prevents the germination (growth) of other plant seeds. The chemical can be extracted from the leaves.

(a) A student carried out an investigation into the effect of this chemical on mung bean seeds. Leaf extracts containing different concentrations of the chemical were prepared.

The student was supplied with

30 mung bean seeds   a bottle of 0·1% leaf extract chemical
3 identical petri dishes   a bottle of 1% leaf extract chemical
cotton wool   a bottle of 10% leaf extract chemical

(i) Complete the diagrams below to show how the investigation should have been set up. Label the contents of each petri dish.

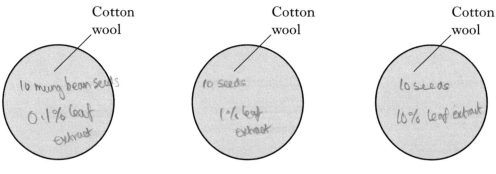

Petri dish 1   Petri dish 2   Petri dish 3

(ii) What observations and measurements should be taken to obtain results for this investigation?

_____

_____

(iii) A control petri dish should also have been set up to show that it was the leaf extract preventing the growth of the mung bean seeds.

Complete the diagram below to show the contents of the control petri dish.

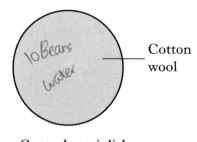

Control petri dish

(b) Explain why producing this chemical is an advantage to the black walnut trees.

_____

_____

11. Charles Darwin visited the Galapagos Islands. He found different species of finch on the different islands.

The following gives information on the size and shape of beaks and the island habitats of two of the Galapagos finches.

Size and shape of beak	Habitat
Long and narrow	Rotting logs that provide food for insects
Short and wide	Trees and shrubs that provide seeds and nuts

Finch A

Finch B

(a) State which finch eats insects and give a reason for your answer.

Finch _____

Reason _____

(b) Identify **two** ways in which competition between finch A and finch B is reduced.

_____

_____

[Turn over for SECTION C on *Page twenty-two*

SECTION C

Both questions in this section should be attempted.

Note that each question contains a choice.

Questions 1 and 2 should be attempted on the blank pages which follow.

Supplementary sheets, if required, may be obtained from the invigilator.

1. Answer **either** A **or** B.

   A. The diagram below shows some characteristics of two present day breeds of dog which descended from a wolf-like common ancestor.

   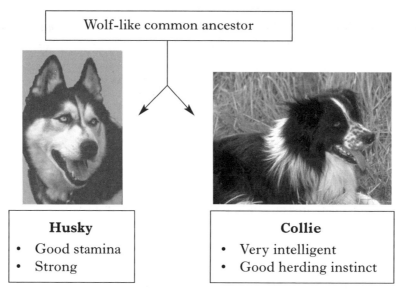

   Name and describe the process which humans have used to produce different breeds of dog.   5

   OR

   B. The diagram below shows the two different forms of the peppered moth *Biston betularia* on the bark of a tree located in an unpolluted area.

   Name and describe the process by which the black form of the moth became the most common form in polluted areas of Scotland.   5

Question 2 is on *Page twenty-four*.

SPACE FOR ANSWER TO QUESTION 1

Marks

2. Answer **either** A **or** B.

   Labelled diagrams may be included where appropriate.

   A. Describe how cells are used in the production of yoghurt and alternative fuel. Include in your answer for both, the type of cell used, the substrates and the products. 5

   OR

   B. The rate of photosynthesis is limited by certain environmental factors.

   Name **two** limiting factors and describe how the growth of greenhouse plants in winter can be increased. 5

[END OF QUESTION PAPER]

SPACE FOR ANSWER TO QUESTION 2

Official SQA Past Papers: Intermediate 2 Biology 2005

SPACE FOR ANSWERS

ADDITIONAL GRAPH PAPER FOR QUESTION 3(b)

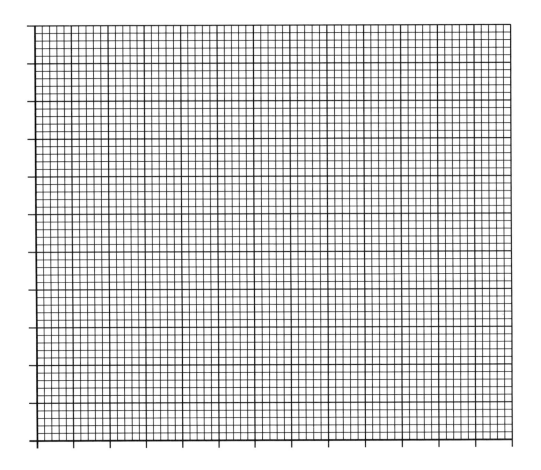

ADDITIONAL SPACE FOR ANSWERS

[BLANK PAGE]

[BLANK PAGE]

[BLANK PAGE]